Technology, Ethics and the Protocols of Modern War

Contemporary security has expanded its meaning, content and structure in response to globalisation and the emergence of greatly improved world-wide communication. The protocols of modern warfare, including targeted killing, enhanced interrogations, mass electronic surveillance and the virtualisation of war have changed the moral landscape and brought diverse new interactions with politics, law, religion, ethics and technology.

This book addresses how and why the nature of security has changed and what this means for the security actors involved and the wider society. Offering a cross-disciplinary perspective on concepts, meanings and categories of security, the book brings together scholars and experts from a range of disciplines including political, military studies and security studies, political economy and international relations. Contributors reflect upon new communication methods, postmodern concepts of warfare, technological determinants and cultural preferences to provide new theoretical and analytical insights into a changing security environment and the protocols of war in the 21st century.

A useful text for scholars and students of security studies, international relations, global governance, international law and ethics, foreign policy, comparative studies and contemporary world history.

Artur Gruszczak is an associate professor of political science, chair of national security at the Department of International and Political Studies, Jagiellonian University in Krakow, Poland. He is also faculty member of the European Academy Online run by the Centre International de Formation Européenne in Nice. His principal interests and research areas include: EU internal security, EU intelligence cooperation, Euro-Atlantic security, modern warfare theory. Recently he published *Intelligence Security in the European Union, Building a Strategic Intelligence Community* (Palgrave Macmillan 2016) and edited *Euro-Atlantic Security Policy: Between NATO Summits in Newport and Warsaw* (Institute of Strategic Studies in Krakow 2015).

Paweł Frankowski is an associate professor of international relations at the Institute of Political Science and International Relations, Jagiellonian University in Krakow, Poland. He was post-doctoral fellow at University of Iceland (2010–2011) and University of St. Gallen (2011–2012). He was also a 2008 US State Department Fellow, SCIEX 2012 Fellow and 2015 Salzburg Global Seminar Fellow. His current research interests include: outer space security, social standards in free tree agreements and regional integration schemes in Africa.

Emerging Technologies, Ethics and International Affairs
Series Editors: Steven Barela, Jai C. Galliott, Avery Plaw,
Katina Michael

This series examines the crucial ethical, legal and public policy questions arising from or exacerbated by the design, development and eventual adoption of new technologies across all related fields, from education and engineering to medicine and military affairs. The books revolve around two key themes:

- Moral issues in research, engineering and design
- Ethical, legal and political/policy issues in the use and regulation of technology

This series encourages submission of cutting-edge research monographs and edited collections with a particular focus on forward-looking ideas concerning innovative or as yet undeveloped technologies. Whilst there is an expectation that authors will be well grounded in philosophy, law or political science, consideration will be given to future-orientated works that cross these disciplinary boundaries. The interdisciplinary nature of the series editorial team offers the best possible examination of works that address the 'ethical, legal and social' implications of emerging technologies.

For more information about this series, please visit: https://www.routledge.com/Emerging-Technologies-Ethics-and-International-Affairs/book-series/ASHSER-1408

Technology, Ethics and the Protocols of Modern War

Edited by
Artur Gruszczak and Paweł Frankowski

LONDON AND NEW YORK

First published 2018
by Routledge
2 Park Square, Milton Park, Abingdon, Oxon OX14 4RN

and by Routledge
605 Third Avenue, New York, NY 10017

First issued in paperback 2021

Routledge is an imprint of the Taylor & Francis Group, an informa business

Publisher's Note
The publisher has gone to great lengths to ensure the quality of this reprint but points out that some imperfections in the original copies may be apparent.

British Library Cataloguing in Publication Data
A catalogue record for this book is available from the British Library

Library of Congress Cataloging in Publication Data
Names: Gruszczak, Artur, 1965- editor of compilation. | Frankowski, Pawel, editor of compilation.
Title: Technology, ethics and the protocols of modern war / edited by Artur Gruszczak and Pawel Frankowski.
Description: Abingdon, Oxon; New York, NY: Routledge, [2018] | Series: Emerging technology, ethics and international affairs series no. 1; Ashser-1408 | Includes bibliographical references and index.
Identifiers: LCCN 2017051109| ISBN 9781138221338 (hardback) | ISBN 9781315410739 (e-book)
Subjects: LCSH: War—History—21st century. | Security sector—History—21st century. | Military art and science—History—21st century. | Military art and science—Technological innovations—Moral and ethical aspects. | Cyberspace operations (Military science) | Security, International.
Classification: LCC U21.2.T365 2018 | DDC 172/.42—dc23LC record available at https://lccn.loc.gov/2017051109

ISBN 13: 978-1-03-209563-9 (pbk)
ISBN 13: 978-1-138-22133-8 (hbk)

Typeset in Times New Roman MT Standard
by diacriTech, Chennai

Table of contents

List of figures

List of tables

Notes on contributors

Isil Akbulut is a PhD candidate in the Department of Political Science at Wayne State University in Detroit, MI. Her main areas of interests are: global governance, peace operations, social network analysis and terrorism. Her dissertation explores inter-organisational collaborations in peace operations.

Joe Burton is a senior lecturer in international relations at the New Zealand Institute for Security and Crime Science, University of Waikato, New Zealand. He has a PhD and Master of international studies degree from the University of Otago and an undergraduate degree in international relations from the University of Wales, Aberystwyth. He has worked in professional politics as a ministerial advisor, national campaign coordinator, legislative assistant, researcher and political organiser. His research is currently focused on cyber security and how states, non-state actors, international organisations and alliances are adapting to deal with new strategic challenges.

Dighton (Mac) Fiddner is an associate professor of political science at Indiana University of Pennsylvania. He holds an MA in political science from Kansas University, and a PhD in political science from the School of Public and International Affairs at University of Pittsburgh. He currently teaches international relations, American foreign policy, intelligence and strategy policy courses as well as the information systems as a national security risk. Prior to his academic career, Dr Fiddner served in the US Army, retiring as a lieutenant colonel in September 1988.

Paweł Frankowski is an associate professor of international relations at the Institute of Political Science and International Relations, Jagiellonian University in Krakow, Poland. He was post-doctoral fellow at University of Iceland (2010-2011) and University of St. Gallen (2011-2012). He also was a 2008 US State Department Fellow, SCIEX 2012 Fellow and 2015 Salzburg Global Seminar Fellow. His current research interests include: outer space security, social standards in free tree agreements and regional integration schemes in Africa.

Crister S. Garrett is professor for American, International and Transatlantic studies at the Universität Leipzig, Germany. He earned his PhD at the University of California Los Angeles (UCLA) and taught at the Middlebury Institute of International Studies and the University of Wisconsin-Madison before assuming his current position. He has been a Fulbright Fellow, a Robert Bosch Fellow and a DAAD Fellow. His research focuses on cultures of security in a comparative perspective.

Artur Gruszczak is an associate professor of political science, chair of national security at the Department of International and Political Studies, Jagiellonian University in Krakow, Poland. He is also faculty member of the European Academy Online run by the Centre International de Formation Européenne in Berlin. His principal interests and research areas include: EU internal security, EU intelligence cooperation, Euro-Atlantic security, modern warfare theory. Recently he published *Intelligence Security in the European Union. Building a Strategic Intelligence Community* (Palgrave Macmillan 2016) and edited *Euro-Atlantic Security Policy: Between NATO Summits in Newport and Warsaw* (Institute of Strategic Studies in Krakow 2015).

Iveta Hlouchova earned her doctorate in political science at the Masaryk University in Brno, Czech Republic. She holds a Master's degree in security and strategic studies from the same university. She also graduated from the George C. Marshall European Center for Security Studies and has become one of the adjunct professors acting as facilitators for the Marshall Center Program on Terrorism and Security Studies. Iveta is also an independent course instructor and security consultant on various aspects of politically motivated violence.

Alex C. Hochuli leads a research project on the transformation of European armed forces at the Military Academy at ETH Zurich, Switzerland with a special focus on transformation in Germany, France and Sweden. He studied political science, sociology and economics at the University of Zurich. His research focuses on security politics in Europe and its political and economic implications. He holds a degree in political science, sociology and economics from the University of Zurich.

Michael O. Holenweger is a lecturer at the ETH Zurich, Switzerland and project leader of a research project at the Military Academy at ETH Zurich that is concerned with leadership in extreme situations. He studied political science, communication studies and ethnology at the University of Zurich and teaches, among other things, courses on the topics of leadership and communication. In addition, he acts as a consultant in politics, particularly in the area of security and foreign policy and as a consultant for crisis management in international companies. He is co-editor of *Leadership in Extreme Situations*, published by Springer (2017).

Muthumanickam Matheswaran has a PhD in defence and strategic studies from University of Madras, India. He is Air Marshal (Retired) of the Indian Armed Forces. He was the Deputy Chief of Integrated Defence Staff. He is a visiting doctoral faculty at the Naval War College at Goa, India. He is a strategic consultant and advisor to defence and aerospace companies. He is an adjunct professor at the Department of Geopolitics, Manipal University, India. He continues to practice as a strategic analyst in national and international security, defence and aerospace and presents papers at various international conferences.

Michał Pawiński obtained PhD in 2017 from Graduate Institute of International Affairs and Strategic Studies, Tamkang University, Taiwan. He has been a recipient of scholarships and grants from the Ministry of Education and Ministry of Science and Technology of the Republic of China in Taiwan. He is the co-author of *People's Liberation Army* published in Polish with the Polish Academy of Sciences in 2014. His articles appeared in *International Peacekeeping, Journal of Military Ethics* and *Strategic Vision*.

Frederic S. Pearson is director of the Center for Peace and Conflict Studies, professor of political science and Gershenson Distinguished Faculty Fellow at Wayne State University, Detroit, MI. He was previously professor and research fellow at the University of Missouri-St. Louis and twice a Fulbright scholar, in the Netherlands and UK. Recent research projects have dealt with inter-governmental organisation collaboration in peacemaking missions and with insurgent group structures, networks and violence levels.

Błażej Sajduk holds a doctoral degree in political science from Jagiellonian University in Krakow, Poland. He is an assistant professor of political science at Jagiellonian University and Tischner European University in Krakow. His research interests focus on social and ethical dimensions of the use of new technologies, political analysis and political philosophy.

Simona R. Soare is an associate researcher at Institut d'Etudes Européennes, l'Université Saint-Louis in Brussels. She holds a PhD in Political Science/International Relations from the National University of Political and Administrative Studies in Bucharest, Romania. She has authored numerous publications on transatlantic security, military capability building programmes and US defence policy.

Andrea Kathryn Talentino is vice president for academic affairs and professor of political science at Nazareth College in Rochester, New York. She was previously professor at Drew University and Tulane University, and a Sawyer Seminar post-doctoral fellow at the Center of International Studies at Princeton University. Recent research projects have dealt with inter-governmental organisation collaboration in peacemaking missions, as well as connections between Olympic sport, peacekeeping and international norms.

Veena Thadani is a clinical associate professor and coordinator of politics in the Paul McGhee Division of the School of Professional Studies at New York University in New York, NY. Her recent research is focused on the contradictions of globalisation, the globalisation of poverty and under-development. Her publications include policy papers and journal articles in the field of third world development drawing on her experience as a research scholar in a population and development think-tank.

Caroline Varin holds a PhD in international relations from the London School of Economics. She is a lecturer in international security at Regent's University London and associate of the Global South Unit, London School of Economics. She has published books on mercenaries with Routledge (2014), on Boko Haram with Praeger (2016) and on Violent Non-State Actors with Palgrave (2017).

1   Introduction

Protocols of modern war

Artur Gruszczak and Paweł Frankowski

Security studies have evolved into a robust, dynamic and forward-looking discipline of the social sciences. This is largely due to the growing importance of security in everyday activity of individuals, social groups, enterprises and states. It is also owing to the development, enlargement and enrichment of theoretical reflection and empirical research on security in the volatile and erratic world of the 21st century. New security challenges which have emerged since the beginning of the present century demand both a premeditated and tenacious political response as well as insightful consideration and afterthought. They reflect and determine the nature and structure of security and its embedding in the complex social, economic, political and cybernetic systems. The increasing complexity of the security environment contributes to the emergence of new unconventional and postmodern risks and threats as well as the rise of unmodern men and societies who reject the global order based on universal values of mankind.

Contemporary security has expanded its meaning, content and structure. It has accompanied the emergence of cyberspace, globalisation of the economy, worldwide communication and cultural diversity. It has entered into diverse interactions with politics, law, culture, religion, ethics and technology. These interactions have often brought about substantial and challenging consequences expressed in different forms, modes and categories. They are interrelated, mutually referencing and self-sustaining, i.e., the variables belonging in a given dimension of security (political, economic, cultural, technological and societal) interact with each other and produce noticeable effects in different realms. For instance, cultural changes affect the political system and social order which can cause serious disturbance of public security and may require normative changes in order to counteract subversion.

This book is a collection of articles looking at contemporary security issues from different angles in order to highlight the complex and diverse nature of today's security and at the same time point to a conceptual 'interconnector' enabling and validating the cross-disciplinary approach adopted in this volume. The so-called 'protocols of war' play the role of such an interconnector stimulating the circulation of ideas, concepts and meanings of security. They also provide an analytical platform for studying strategic, political, technological and ethical aspects of contemporary security. They frame the applied concepts of warfare and also underpin a strategy of conflict resolution and crisis management in social, cultural, technological, ethical and legal contexts. The protocols of war opened in the 21st century ('War on Terror', 'War for Iraqi Freedom', 'Counterinsurgency Warfare', 'Hybrid War', 'Drone Wars') have perverted the rugged moral landscape drawn by the 'good old' wars of the 20th century. Targeted killing, enhanced interrogations, human terrain system, and also mass electronic surveillance, virtualisation of war, post-truth public discourse and internet trolling are epitomes of demoralisation of force and tolerance for its indiscriminate use. Hyperactivity of unconventional actors, such as suicide bombers, child soldiers, cyber squads, private military contractors and paramilitary proxy groups ('little green men') (Plaw and Augé 2012, p. ix) has fuelled unprecedented outbursts of anger, aggression, rage and cruelty which were transformed into violent rituals as a highly coordinated form of collective violence (Tilly 2003, pp. 24–26).

We assume that a rationale for the increasingly indiscriminate infliction of violence bears profound moral, ethical and legal consequences for the metamorphoses of modern war. Technology as a 'power generator' enhances the blurred picture of contemporary security in terms of ethics, justice and moral reasons. The development of complex autonomous security systems may bring about positive effects in terms of technical effectiveness, controllability and reliability yet it may also lead to safety-critical situations and cause serious harm. In the world of 'killer applications' (Singer 2010) ethical issues and moral dilemmas too often face the onslaught of strategic imperatives, modi operandi, military necessity, command responsibility, etc. Deception and manipulation, enhanced by mass communication technologies and cyber methods, have become offensive harmful forms of confrontation practiced by anonymous entities posing normative dilemmas and legal challenges for domestic law and international law (Boothby 2014, pp. 8–9).

By introducing the concept of the protocols of war, we attempt to challenge conventional normative and explanatory categories underpinning observational rationality in peace and conflict studies. Confusion caused by observation and analysis of so many complex processes and phenomena in the contemporary security environment, which often overlap themselves and contradict each other, must not undermine the cognitive flexibility required for a comprehensive approach to modern war. Technological and cultural production of violence (Whitehead 2007) breeds multiple concerns when it comes both to its immediate effects and long-term consequences of 'collateral moral damage'. It makes us ponder over what feeds violent behaviour and what limits or restraints can be imposed on its contagious outcomes, especially in social networks and state institutions. We argue that 'protocolarisation' of war and violence is a rational-choice response to growing complexity and ambiguity of contemporary social, economic and political systems whose agents and structures strive to avoid or escape anarchical tendencies. We believe that the perpetual interaction between rationality and emotions has been predetermined by normative principles and regulatory guidelines which shape individual beliefs and basic preferences of actors. In contemporary complex and precarious systems, strategic actors attempt to reduce the complexity of the system and its interdependencies. This is particularly relevant in conflict situations and during wartime. Violence and military force used to be applied either premediately or impetuously, subject to interplay between strategic and tactical levels of hostilities. Political decision makers, military commanders or warlords tend to reduce the amount of information needed for managing the area of confrontation and make use of simplified procedural schemes. In pursuit of an effective way of closing the complexity gap between warfare and military organisation along with its steering mechanisms, protocols of war are introduced in order to reduce pressure on the apparatus of political decision making, streamline the command chain and mitigate moral and cultural harm-based restraints (Rose 2011, chap. 5). As to the latter, the protocols of war help to handle the 'impossibility theorem' characteristic for cultural theory of conflict (Chai and Wildavsky 1994, p. 163).

Our concept of the protocol of war departs from narrow terms typical for a technical notion of the word 'protocol'. Dictionary definitions of protocol point to rules as the core part of this term and formal modes of their implementation. Merriam-Webster (2016) proposes the most general definiton: 'a system of rules that explain the correct conduct and procedures to be followed in formal situations'. Other expressions point to specific usage of the word 'protocol' in diplomacy, social relations and technology. For instance, Webster's New World Dictionary (Guralnik 1986, p. 1143) provides a narrow definition proper for computer science: 'set of rules governing the communication and the transfer of data, which is particularly proper for computer network systems'.

Building on Alexander Galloway's view, according to which protocol refers to 'standards governing the implementation of specific technologies' (Galloway 2004, p. 7), we see the concept of the protocols of war particularly applicable to sensitive structures and complex frames of multi-layered contemporary forms of organised violence. In a nutshell, *the protocol of war is a codified system of values, norms and rules that legitimise and explain forms, methods and means of violence in domestic and international affairs*. It is also a recognised practice-by-experience accompanying the evolution of warfare and the development of the art of war. What matters for the present book is the cognitive value of the notion of the protocol of war lying in its capability to interconnect security components located at different levels or embedded in different structures.

Although Galloway (2004, p. 29) claims that protocol in the military context is seen as 'a method of correct behavior under a given chain of command', we do not conceive of protocols of war as 'codes of warfare', sets of formalised procedures, actions and practices, enshrined in international conventions and protocols defining the legal conditions of use of force. Protocols are about management, modulation and control. They are as concerned with disconnection as they are with connectivity (Thacker 2004, pp. xvii–xviii).

The protocols of war by no means can be identified with *topoi* of war, archetypical narratives describing violence and killings, deeply embedded in the historical memory and reproduced culturally by future generations of ethnic, religious or national communities. Rather, they reflect changeability, fluidness and ambivalence of the contemporary world where the ways and means of social and political becoming are interchangeable and self-referential. Therefore, any *topos* of modern war is doomed to obfuscate and distort reality because violence, force and outrage are deeply nested in frames of reference specific for a given stage of human development and organisation. In a nutshell, modern war has been conditioned by a complex set of ideas, rites and narratives which were deliberately fostered and extensively enhanced by technologies, ideologies and morals. Vaccari (2015, p. 21), drawing on Deleuze and Guattari's nomadology, ascertains that: 'The war machine brings its own patterns of thought, codes and ways of organising and occupying space'. We argue that modern war generates new sets of codes, rules and norms responding to structural dynamics of conflict, coercion and violence.

Moreover, following Bialasiewicz et al. (2007, pp. 407–409; comp. Butler 2009, pp. 7–9) we claim that the protocols of war introduce discursive practices aiming to articulate specifications of violence through a combination of recognised ethical, technological and organisational patterns which provide a mechanism of self-sustainment of conflict-prone dispositions and forms of behaviour. It is worth underlining that the emergence of unconventional actors, as well as new technologies and devices applied in military confrontation, generated new performative discourses (Bialasiewicz et al., 2007, pp. 406–407) labelling different varieties of violence and embedding them in popular consciousness through the use of mass communication techniques. As a result, the protocols of war entered into circulation as frames of reference helping to understand and interpret 'war messages'. Consequently, they were used to seek publicity or provide rationale for unconventional (asymmetric, hybrid, non-linear) modes of warfare practiced by amorphous warring subjects. Bialasiewicz et al. (2007) as well as Sylvia (2013, pp. 3–4) use the term 'codes of security' and point to the post-9/11 'global war on terror' as a classic example of circulation of performative discourses and 'branding messages'.

The aim of this collection is to utilise the concept of the protocols of war for responding to the two following questions: (1) why and how has the nature of security changed so extensively and (2) what are the most remarkable implications of the accelerated evolution of contemporary meanings and forms of security? So, the book seeks to provide a comprehensive analysis of the dynamic of security structures and systems permeated by the protocols of war. The goal of the editors of this volume is to move a reflection on the changing nature of contemporary security to a deeper level of cross-disciplinary analysis enabling exploration of strategic, technological, ethical and societal aspects. The contributors in this volume take different positions and try to broaden their cognitive perspectives, seeking to highlight diversity rather than pursuing narrow paths of conventional security studies. Some directly refer to certain features of security, such as cyberspace, insurgency, use of drones or strategic communication. Others offer more theoretical and reflexive approaches to broadly-viewed security emphasising such aspects as social justice, militarisation and post-conflict reconstruction.

The book offers a cross-disciplinary perspective on concepts, meanings and categories of security. The cross-disciplinarity of this collection is additionally highlighted by the fact that the book brings together a group of scholars and experts representing different traditions, schools, generations, institutions and scientific communities from all over the world. Editors and authors represent a wide range of disciplines such as military studies, security studies, international relations, International Political Economy and contemporary studies. The chapters examine links between military practice and academic overlook, security and peace studies. Such a variety of views, approaches and opinions gives an additional flavour to the whole collection, enriching the analytical value and conceptual novelty of the book. It aims to underscore the changing perspectives on security, subject to normative, cultural, technological and political variables. Several contributions highlight the meaning of power in contemporary security studies and point to the problem of violence and indiscriminate use of force. Other texts explore the impact of new technologies, particularly the effects of virtualisation of the security arena and their relevance for understanding security in different contexts. Upon closer consideration, one can notice a plethora of questions and issues that are reconsidered and subjected to an intense scrutiny. If they still create an impression of conceptual eclecticism, it could be somehow defended by reference to the methodological paradigm of analytic eclecticism introduced to the social sciences, and specifically to IR studies, by Sil and Katzenstein (2010).

In this book, editors and authors point out that technology has exerted a particularly strong impact on today's security. The emergence of virtual reality, the global expansion of the Internet and cellular telephony, the rise of robotics and the possibility to do remote jobs heralded a 'brave new era'. It had pride in progressive expansion of high-tech devices and solutions, user-friendly global communication and enormous, extraordinary virtual and physical mobility of people and goods. On the other hand, technology proved to be a double-edged weapon, enabling the proliferation of hate, fanatism and extermism as well as facilitating illegal circulation of money, weapons and drugs, conceived as traditional threats for security. Military technologies have exerted powerful impact on security culture and organisation. It can be observed not only from the perspective of evolution of arms, military equipment and combat systems, but should be analysed in the much wider context of the evolution of contemporary war. The massive use of drones is one of the most glaring current examples of how technology determines the 'way of warfare'. One should also take into account cultural and educational repercussions of security-oriented public policies and social behaviours. Public discourse has increasingly displayed a military- and security-oriented bias, often adjusted to the prevailing collective mindsets.

The book is divided in three parts. The first part addresses security in its dynamic, changing meaning, pointing to evolution and transformation of the organised forms of violence. The second part focuses on technological aspects, underlining the expansion of technologies, the rise of remotely-controlled devices, the dilemma of dual-use technologies and the impact of military technologies on education and science. The third part elaborates on current security dilemmas, focusing on terrorism and cyber warfare as the major global threats.

Part I begins with an analysis of strategic communication and its role in a complex information environment. Aware of the communication context of the protocols of war, Alex Hochuli and Michael Holenweger examine NATO standards of strategic communication and their impact on the allied concept and doctrine as well as responses from several member states. The dynamic of the complex security environment is assessed and analysed in a new conceptual context by Artur Gruszczak. He offers a critical assessment of post-Cold War concepts of violence and warfare, especially the theories of asymmetric conflicts and hybrid wars. He puts forward a concept of postmodern warfare as an analytical tool useful for the interpretation of contemporary protocols of war highlighting non-optimal effects of diversity and a binary approach to norms and values.

Security has been more often regarded not as a value or an ingredient of the normative system, but as a commodity provided by different stakeholders active on the global market. It has been offered by the increasing number of private security providers marking their presence not only in public facilities, transportation or military logistic and training but also at the frontline of armed conflicts. The expanding area of security outsourcing and state-licensed violence has raised moral issues and ethical dilemmas. The issue of legitimisation of violence as commodity is taken up by Iveta Hlouchova. She juxtaposes private security and military companies with foreign fighters as entities operating in the same theater of conflict which are neither isolated, nor independent. We can say that both actors write up the same protocol of war although each of them reads out, interprets and understands it in different manners. Moreover, Hlouchova underscores opposition between private military contractors and foreign fighters and focuses on the former as strategic player countering the phenomenon of foreign fighters. The market dimension of violence is investigated by Caroline Varin. She stems from the linkage between inequality and political instability and conflict around the world. She claims that inequality was a mobiliser for uprisings in the Arab Spring and across Africa, and suggests that the only way to address this issue is to securitise inequality and develop policies that effectively handle this matter. The nature of conflicts analysed by Varin can be partially noticed in the case of Ukraine presented by Crister S. Garrett. He points to cultural explanations of the turmoil in Ukraine and puts forward the concept of the 'ecosystem of international security'. He elaborates on a dynamic cultural model that frames narratives of security. Garrett's grasping of cultural complexity of contemporary conflicts corresponds with the meaning of protocol as framing production of referential schemes according to adopted rules (see Goffman 1986; Butler 2009).

Part II is devoted to the technological and ethical aspect of contemporary conflicts and rivalry. Muthumanikam Matheswaran builds his contribution on the concept of global commons arguing that solutions to security challenges lie in adopting a globally common, interdisciplinary approach which balances a state-centric approach. Matheswaran emphasises the growing relevance of digital technologies accelerating the process of globalisation and interdependency. Given that globalisation implies deterritorialisation, he advocates for an interdisciplinary approach to security that better reflects complex interdependence between technology, economy, military power and trade.

Global aspects of international peace and peacemaking are analysed by Andrea Kathryn Talentino, Frederic S. Pearson and Isil Akbulut. They focus on interorganisational relations in humanitarian interventions as particularly important in disarmament and demobilisation efforts. They argue that success in disarmament demands extensive collaboration among involved local and international actors across a wide range of contexts within the target country. They use the cases of Liberia and Sierra Leone for analysis of disarmament strategies proposed or implemented by different governmental and nongovernmental organisations.

Warfare in human affairs sometimes entails specialty and extraordinariness. This also refers to technology and science in experimental and research phases. The Tuscaloosa Experiment or the MK-Ultra and Phoenix programmes, analysed by Michał Pawiński in this volume, illustrate the importance of interdisciplinary research for military purposes. Pawiński warns that the involvement of social scientists in military operations and intelligence collection might not be ethically and morally acceptable. The historical case of MK-Ultra and the contemporary Human Terrain System project provide evidence for militarisation of social sciences along with far-reaching consequences for engagement of academia in military-driven projects and possible abuses of cooperation between the army and academia.

The impact of the military on research and technology development is further elaborated in the next two chapters. Simona R. Soare analyses the trends in European military technology transfers to Russia and reflects on their consequences for the relationship between the EU and Russia as well as European defence priorities. Building on the analysis of trends in arms transfers, the chapter raises political-military, legal, economic and security aspects of the European military transfers to Russia. Military technology advancements and their ethical consequences are brought forward by Błażej Sajduk. In his contribution, the paradigm of post-heroic warfare is introduced as a reference for the advent of autonomous systems implying the growing presence of artificial intelligence in military operations. The consequences of the development of new advanced autonomous weapon systems are particularly remarkable and sensitive for moral and ethical concerns. Sajduk convincingly reports the passage from the traditional military code of conduct to a modern protocol of dehumanised warfare guided by algorithms and 'de-ethicised' commands.

Part III of the book takes up the issues of counterterrorism and cyber security. Veena Thadani provides an impressive picture of local and indigenous insurgencies fed up by unresolved dilemmas of internal security and social justice. She examines the Naxalite insurgency in eastern and central India, focusing on the spread of the Maoist insurgency and its connection to the failure of the development agenda proposed by the Indian state in its policies of land seizures and development-induced displacements in the mineral-rich areas. Thadani outlines the protocol of 'develop or perish' and highlights its failure in addressing demands for social and economic justice.

The last two chapters address the issue of cyber security and defence. Joe Burton examines the European Union Cyber Security Strategy and the NATO Cyber Defence Policy, focusing on the emergence of cyber security norms through their respective policy frameworks. He argues that cyber security norms are at an early yet promising stage of development in the Euro-Atlantic area. He asserts that closing the gap between academics, policymakers, computer scientists and the private sector is essential in enhancing cyber security and embedding cyber security norms within the international system. While Burton seems to take cyberspace as an increasingly relevant though auxiliary dimension of security, Mac Fiddner asserts that nowadays cyberspace has become central to international security. He elaborates on ontological aspects of cyberspace claiming that currently it produces changing effects on security threats and response vectors by diffusing the sphere from where threats emanate and the possible response to them. He addresses specific risks within the cyberspace domain and refers them to the broader strategic environment.

This book makes a new contribution to the ongoing debate on contemporary security in the rapidly evolving local and global environment. It seeks to provide a comprehensive analysis of the dynamics of security structures and systems permeated by the protocols of war. The goal of the editors of this volume is to move the reflection on the changing nature of contemporary security to a deeper level of cross-disciplinary analysis enabling exploration of strategic, technological, ethical and societal aspects. The contributors in this volume take different positions and try to broaden their cognitive perspectives, seeking to highlight diversity rather than pursuing narrow paths of conventional security studies. Upon closer consideration, one can notice a plethora of questions and issues that are reconsidered and subjected to an intense scrutiny. The dynamics of political, social, economic, and other systems, liquidity of material goods and intangible assets as well as relativity of norms and values have contributed to the fast changes of security landscapes stimulated by the moving contexts of the actors' performance. The roles and games played by security actors fill the arena of contemporary security with evolving approaches, autonomous actions, critical postures and shifting identities. The texts included in the proposed collection reflect upon this extremely complex and subtle fabric of security. They provide theoretical and analytical insights into the latest and most relevant features of security transformations in the 21ˢᵗ century, focusing on new communication and narratives, postmodern concepts of warfare, technological determinants, normative frameworks and cultural preferences, diplomacy and cyberspace.

This volume contains revised versions of 13 chapters deliberately selected for publication from almost 150 papers presented at the Contemporary Interdisciplinary Security Studies Conference held in June 2015 in Krakow, Poland.

References

Bialasiewicz, L. et al. 2007, Performing security: The imaginative geographies of current US strategy, *Political Geography*, 26(4): 405–422.

Boothby, W.H. 2014, *Conflict Law. The Influence of New Weapons Technology, Human Rights and Emerging Actors*. The Hague, the Netherlands: T.M.C. Asser Press.

Butler, J. 2009, *Frames of War. When Is Life Grievable?* London and New York, NY: Verso.

Chai, S.-K. and Wildavsky, A. 1994, 'Culture, Rationality, and Violence', In D.J. Coyle and R.J. Ellis (eds), *Politics, Policy, and Culture*. Boulder, CO: Westview Press, 159–174.

Galloway, A.R. 2004, *Protocol: how control exists after decentralization*. Cambridge, MA and London: The MIT Press.

Goffman, E. 1986, *Frame Analysis. An Essay on the Organization of Experience*. Boston, MA: Northeastern University Press.

Guralnik, D. B. (ed). 1986, *Webster's New World Dictionary of the American Language*. New York, NY: Prentice Hall Press.

Merriam-Webster Dictionary. 2016, 'protocol', viewed 20 September 2016, www.merriam-webster.com/dictionary/protocol.

Plaw, A. and Augé, A. 2012, 'Introduction: The Transformations of War', In A. Plaw (ed), *The Metamorphosis of War*. Amsterdam, the Netherlands and New York, NY: Rodopi, 9–18.

Rose, D.C. 2011, *The Moral Foundation of Economic Behavior*. New York, NY: Oxford University Press.

Sil, R. and Katzenstein, P.J. 2010, *Beyond Paradigms: Analytic Eclecticism in the Study of World Politics*. Basingstoke: Palgrave Macmillan.

Singer, P.W. 2010, The ethics of killer applications: Why is it so hard to talk about morality when it comes to new military technology? *Journal of Military Ethics*, 9(4): 299–312.

Sylvia, P. 2013, *Cultural Messaging in the U.S. War on Terrorism. A Performative Approach to Security*. El Paso, TX: LFB Scholarly Publishing.

Thacker, E. 2004, 'Foreword: Protocol Is as Protocol Does', In A.R. Galloway (ed), *Protocol: How Control Exists After Decentralization*. Cambridge, MA and London: The MIT Press, xi–xxii.

Tilly, Ch. 2003, *The Politics of Collective Violence*. Cambridge and New York, NY: Cambridge University Press.

Vaccari, A. 2015, 'Abjecting Humanity: Dehumanising and Post-humanising the Military', In J. Galliott and M. Lotz (eds), *Super Soldiers. The Ethical, Legal and Social Implications*. Farnham and Burlington, VT: Ashgate, 9–24.

Whitehead, N.L. 2007, Violence and the cultural order, *Daedalus*, 136(1): 40–50.

2 Strategic communication and contemporary European security

Michael O. Holenweger and Alexandre C. Hochuli

Introduction

We live in mediated times. While billions are reached more or less indirectly through mass media channels which still rest on the advent of print innovations from the 18th century, more and more are reached directly online over social media and through self-made broadcasting systems. The sender and receiver of media content have grown closer like never before. In this increasingly interconnected system, even rebels and insurgents use public diplomacy and social media to reach out to adversaries and allies alike (Huang 2016). It is such an environment in which states find themselves as they try to pursue their strategic interests. Communication, therefore, has been playing an increasingly important role in state security.

It is no longer possible for state governments to ignore reactions to their policy agenda. As the cases of *Wikileaks* and *Panama Papers* have shown, inaction to global news revelations can be devastating to the political establishment. This became particularly true in the case of military contingency operations of the West in the past decade, with operations in Afghanistan from 2001 and in Iraq from 2003 onwards as a case in point. Increasingly, states found themselves locked in a vexing challenge of reacting to the media activity of foreign and domestic actors, including public activity of insurgent networks. Aspects such as selling efforts and achievements on the ground to local and home audiences became crucial to the overall effort. The projection of power, i.e., sending military forces to remote places from the home base in an age of an increasingly skeptical home audience have created big challenges to governments in convincing the home public of the effort undertaken. But communication also evolved on the battle field with armed forces increasingly relying on psychological operations, information campaigns and military diplomacy. Recently, North Atlantic Treaty Organisation (NATO) member states tried to combine all official communication activity of the operational realm under one roof, Strategic Communications (StratCom).

This chapter seeks to underpin the logic of this action, the anatomy of StratCom and the challenges ahead. It points to the fact that armed forces mistakenly relied on StratCom as a weapon. Instead, communicative efforts in complex military operations underwent different processes and influence behaviour than what armed forces expected (Jackson 2016). We apply a neoclassical realist framework of civil-military relations to the two cases of Germany and France, both of which are NATO member countries and have officially adopted NATO StratCom principles, but with different outcomes. While Germany's civil-military relations hampered the development of StratCom, France's more balanced civil-military relations have favoured a StratCom effort that closely resembles US efforts. The approach taken here points to the importance of the shaping power of internal processes of civil-military relations.

Few time periods have seen so many changes in military policies like the ones witnessed in the past decade. Dating back to the thinking about the incorporation of new technological improvements into battle space superiority in the 1980s, the US introduced a concept labeled *Transformation of the Armed Forces* in 1999. In order to build on the improvements of the information and digital revolution, so the story went, revolutionary organisational and procedural innovations were necessary. Old procedures in military development were regarded as useless given the ever-changing security environment after the end of the Cold War. In 2002, at the NATO summit in Prague, the heads of member states agreed to follow this transformation track.

Military operations in Afghanistan and Iraq, however, proved that the subsequent concepts introduced in NATO's armed forces were mismatched with operational reality (Strachan 2010). In particular, the communication efforts of NATO in Afghanistan and the multinational military alliance in Iraq, vis-à-vis the growing insurgency but also the local population, often ended in deadlock or even to the disadvantage of the sender. Therefore, in 2007, the US introduced a concept of StratCom which aimed at streamlining communicative efforts. The concept was subsequently introduced into NATO and all member states agreed on following the conceptual precepts. However, until today, no NATO country has lived up to the task in reaching full implementation of StratCom. While some have made substantial changes to the initial concept, others have failed completely in incorporating it. The reason for this lies in the peculiarities of today's civil-military relations of each member state of the transatlantic alliance. In order to understand the challenges in a communicative unity of effort, we need to understand the complexities of civil-military relations of NATO's member states.

A civil-military relations theory of strategic communications

The initial question of which factors account for the development and introduction of NATO's StratCom strategy leads us, first, to a working definition of military concepts. Under military concepts we understand formalised ideas about the use of armed forces in either peace time or conflict. They provide information and guidance in a three-dimensional way about when, how and to what extent armed forces are deployed to tackle their challenges.

When describes the threshold of deployment, i.e., at what level of escalation [...] of a given crisis is a particular concept is utilised. A concept of disaster management, for instance, is usually expected to be utilised on a lower stage of crisis escalation than the use of nuclear weapons against an adversary.

How describes the use of military resources. Typically, herein lies what we term classic military doctrine, even if military doctrines cover all three dimensions of military concepts discussed here (Høiback 2013).

The *extent* of resources being used is characterised as the share of the armed forces utilised for a given task. With a concept for classic territorial defence, all service branches and almost all units are expected to be deployed. In contrast, for the capability of strategic aerial reconnaissance, only aviation units would be employed.

Further on, military concepts do not originate in an isolated environment of individual ideas of military leaders. In liberal democracies, both civilian and military institutions and stakeholders are involved in an iterative process of armed forces development. It is this process which we call civil-military relations.

Civil-military relations as such are characterised by several factors. In liberal democracies, the primacy of all civilian actors involved, what we label the *civilian sphere*, enjoys decision-making superiority over all representatives of the *military sphere* in structural and substantial terms over how and when the armed forces will be deployed. These decisions shall not solely be regarded as best-case. Although decisions can also turn into a disadvantageous state for the military, the civilian primacy still endures.

In addition, outcomes of civil-military relations are far from rigid but underlie constant momentum of change. The reason herein lies in different aspects. For example, both spheres do not form a monolithic bloc. The civilian sphere consists of at least the executive and the legislative powers, two components which, depending on party preferences and the government system, might take completely opposing sides. The military sphere usually consists of the different service branches which often share differing views on the implementation of the armed forces, yet they share a dependence on financial budget allocations (Zisk 1993). One crucial trend of the preceding 15 years has been the consolidation of service branches under unified commands in order to enable joint operations. This trend has reinforced the existing state of competition between the services for future funding resources and procurement programmes. Although the literature on military innovation does not agree on the causal relation between competitive service branches and military development, the extent of this competition has not declined during this time (Adamsky and Bjerga 2012; Grissom 2007).

Military concepts simply do not solely originate in one sphere or the other. No concept develops completely through civilian intervention or military thinking. Therefore, civil-military relations rest on structural, i.e., relational, and substantial, i.e., functional, processes that change over time. A state and its subordinate institutions, of which the civilian and the military sphere represent a crucial part, is influenced and shaped by external and internal factors. Just as states stand to each other in an interdependent web of interests and incentives while they pursue their security goals, the same goes for their armed forces. The military sphere does not persevere in a strictly national environment but instead is also shaped by the behaviour of other armed forces. Often, military concepts that have proven successful on the battlefield in the eyes of the beholder end up being emulated by other states (Dyson 2012). The adoption of the American-led transformation blueprint of the armed forces serves as a good example of military emulation. Another example is the air power doctrine of the US in the Gulf War of 1991 which, in its aftermath, spread in most Western armed forces. Furthermore, both spheres can be divided into two levels of leverage on the development of military concepts.

On the international level, states are confronted with challenges which originate from other states' behaviour and the security policy environment. In this setting, states pursue self-oriented interests in bilateral and multilateral interactions. The structural and institutional characteristics of a political system hereby specify the ways in which a state can react to challenges. Under structural characteristics, we understand conditions of economic performance and the size of the state alike. For instance, Germany is the largest economy and home to the largest population in Europe. Therefore, Germany structurally shapes the political and economic development of the continent.

Under institutional characteristics, we understand the embedding of a state in international political organisations. Following this logic, Germany, for instance, provides the largest financial share of the European Union, with the highest number of MEPs, and exerts strong influence on the policies of the European Central Bank (ECB). Therefore, we consider Germany as internationally influential.

Regarding the military, institutional factors are also at play in the development process. The US bears more than 22 per cent of the NATO budget and therefore exerts great power on the partners of the transatlantic alliance (NATO 2017).

The resulting domination in terms of NATO's strategic alignment has led to a situation in which American military concepts have been adopted by European partners more easily. NATO therefore captures a central position in the diffusion of military concepts in European armed forces (Deni 2016; Mayer 2014). According to this logic, a growing internationalisation of military development has been taking place in Europe along American standards, albeit to different extents (English 2004).

On the national level, again, both spheres are exposed to several independent factors. In the political sphere, the national political system describes the relationship under which political decisions concerning the military develop. Depending on the center of gravity in this system – presidential, semi-parliamentary or parliamentary – the decision-making momentum inclines to either the executive branch, as is the case in France, or to the parliament, as is the case in Sweden or Germany. Furthermore, depending on the relationship between the executive and the legislative branch, institutional actors, such as defence ministers, can act as innovators or mavericks in defining new military concepts or act as an oversight actor, such as a parliamentary defence commission, in changing or preventing existing concepts (Hochuli 2016). With this in mind, the more the executive rests in a position of strong control by the legislative branch, the more military concepts will be changed in early stages (Wiesner 2010, 2013).

Last but not least, domestic factors of the military also exert power on military concepts. Put simply, we subsume internal institutional relationships of the armed forces under the label of military bureaucracy. There exists a wide variety of cross-case independence of the military bureaucracy. For instance, the Scandinavian model has traditionally given greater autonomy to administrative bodies in their decision-making freedom. In Sweden, up until 2005, this meant total budget control for every service branch of the military. Independent service branches possess comparatively more leverage in influencing military concepts than less independent branches.

While factors on the international level tend to define, introduce and facilitate military concepts, we expect factors on the national level to exert change during the implementation process. The model depicted in Table 2.1 therefore allows consideration of the origins and development of military concepts (Farrell and Rynning 2010; Terriff et al. 2010).

So far, studies have often neglected internal processes of civil-military relations and mainly focused on sociological approaches and the primacy of the political sphere, or put differently, on civilian control of the military. According to Michael Desch, a pioneer of civil-military relations theory, civilian control does not work when the military sphere regularly interferes in decision-making (Desch 1998). This restriction, however, overestimates the use of harmonious relations and the importance of political leadership. Military response does not necessarily lead to a conflictive stalemate. Armed forces can also serve as an impetus for a positive relationship (Janowitz 1964). Desch's approach focuses too narrowly on pure structural positions of both spheres and does not incorporate changing connections between them (Burk 1998, 2002).

Internal military procedures and their connections to civil stakeholders can therefore best be grasped with organisational theories. According to the

Table 2.1 Civil-military relations and factor levels for military change

		Sphere	
		Political sphere	*Military sphere*
Level	international level	Strategic and political environment	Military emulation
	national level	National political scene	Military bureaucracy

majority of the military innovation literature, there is a connection between both internal processes of the political and military spheres (Posen 1984; Rosen 1988). New concepts can therefore be defined and changed in both spheres simultaneously. For instance, concepts can originate in the military sphere, but they need support from civilian actors who allow change. The Soviet response to Western doctrines of the 1960s was primarily shaped by the connection of a crucial set of actors in the Ministry of Defence and the military, both of which pushed for the introduction of new concepts. Decisive for this outcome was the institutional change in Soviet civil-military relations and mavericks, i.e., powerful individual pioneers, on both sides to facilitate change (Zisk 1993, p. 184; Avant 1993; Farrell 2002; Avant 2007).[1] A more recent example of networks between both spheres and doctrinal development can be found in the introduction of the *Counterinsurgency Field Manual* in the US Army and Marine Corps under the leadership of a handful of officers and civilian defense staff. Without the unrelenting input of Lieutenant General David Petraeus, the introduction of the new doctrine would not have been possible in such a short time (Kaplan 2013). Petraeus even managed to introduce a military concept, which guided the political leadership under President George W. Bush and later Barack Obama, to speak of a new strategy in Iraq and Afghanistan under the standards of the COIN doctrine (Strachan 2010). Mavericks often struggle with an organisational culture that reacts to profound change with skepticism or even resistance (Perlmutter and LeoGrande 1982; Herspring 1990; Herspring 1996; Farrell 2001; Terriff 2006; Herspring 2011).

The result of such connections between both spheres, or between interactive and iterative civil-military relations, are mutual expectations of both spheres towards each other. These expectations are twofold. First, relational expectations arise over the question of the proper form of interaction of both spheres. The political sphere expects the military to act according to standards set by the former. In return, the military sphere expects from the political sphere clear political guidance. One classic way to illustrate this expectation was introduced by Peter Feaver who considered relational expectations through agency theory. In Feaver's view, both spheres act rationally and self-interested (Feaver 1999, 2003). This approach manages to address the interactive processes between the policy and the military branch (Feaver 1998). The problem with Feaver is the fact that he treats both sides as monolithic blocs. Reality shows, however, that while the political sphere consists of at least two parts, the executive and the legislative, the military usually also incorporates different service branches with separate ideas and interests. Multiple principals and agents can mislead the causal correlation.

Second, functional expectations arise from the appropriate use of military force. The political sphere expects military expertise regarding the use of force as the military leadership is expected to provide solutions to security challenges and prepare accordingly. On the other hand, the military sphere expects clear strategic coherency about what kind of tasks the military must fulfill and with what means challenges are to be mastered. If at least one of the two expectations get violated on a regular basis, then we expect to see a civil-military conflict, be it of either relational or functional meaning. Such a conflict falls into the relational category when one sphere is acting wrong from the perspective of the other. For instance, members of the political sphere can publicly discredit military conduct, prevent the promotion of top-level ranks or even openly dismiss them. On the other side, the military can use pillory over political behaviour, foot-dragging or lag-rolling (Groitl 2015). A civil-military conflict can fall into the functional category when there is a persistent mismatch over the use of force. The task spectrum of armed forces and their doctrines have changed markedly since the end of the Cold War. In particular, substantial change occurred from territorial defense to the projection of power in low-intensity conflicts with tasks assigned to armed forces which were traditionally conducted by civil actors of non-governmental organisations and the like.

The neoclassical realist premise of civil-military relations enables the conjunction of the international and the national level of our analysis and goes further as classic sociological theories of the past. Considering only one level would overestimate the outcome. Studies which mainly focus on the international, systemic level fail to grasp the subsequent processes on the national level. Studies which mainly focus on the national, sub-systemic level, however, fail to sufficiently grasp the interaction between politics and the military.

Our approach matches external factors as drivers of state behaviour with internal factors as shapers of developments and outcomes. In this sense, security policies, which form a port of military concepts, originate on the systemic level (international level in our model) but only evolve into tangible outcomes on the subsystemic level (national level in our model) (Dyson 2008). Since, to date, no neoclassical realist studies exist which consider the implementation of military concepts, this chapter tries to fill this gap. Drawing on a neoclassical realist approach, we will incorporate the institutional proportions and processes mentioned above in our model (Groitl 2015).

The development of NATO's strategic communications concept

In 2003, one of the shortcomings of NATO's command of the International Security Assistance Force in Afghanistan (ISAF) was the ineffective use of strategic and operational communication to key audiences both at home and in-theatre. For instance, efforts to raise the awareness among the Afghan population that the majority of civilian casualties in Afghanistan did not result from NATO troops but from anti-government forces, i.e., Taliban and Al-Qaida, were almost always thwarted by counter-communication efforts from the Taliban (e.g., portraying NATO troops as foreign occupiers). When ISAF Headquarters handed over security for the whole of Afghanistan in 2003, the Taliban were quick to tell the press that they were to open a liaison office in the Qatari capital of Doha (BBC 2013). President Hamid Karzai was upset about US and Qatari efforts to even consider talks with the Taliban and refused to talk about the signing of a bilateral security agreement with the US. The media victory was the Taliban's. In order to tackle these communication issues, the US Department of Defense introduced the concept of StratCom in 2007 for Afghanistan (Department of Defense 2007). Two years later, NATO emulated this concept when the heads of states and governments drafted a declaration at the summit meeting of the North Atlantic Council in Strasbourg–Kehl in April 2009. The document stated that:

> it is increasingly important that the Alliance communicates in an appropriate, timely, accurate and responsive manner on its evolving roles, objectives and missions. StratCom are an integral part of our efforts to achieve the Alliance's political and military objectives. [...] We underscore our commitment to support further improvement of our StratCom by the time of our next Summit (NATO 2009a, pp. 2–3, paragraphs 4–9; NATO 2009b).

The concept of strategic communications

With respect to the basic concept of StratCom, the relevant literature often refers to Hallahan et al. who define it as 'the purposeful use of communication by an organization to fulfill its mission' (Hallahan et al. 2007, p. 3). Taking this definition as a starting point, Holtzhausen and Zerfass extend the meaning of this term as follows: 'Strategic communication is the practice of deliberate and purposive communication a communication agent enacts in the public sphere on behalf of a communicative entity to reach set goals' (Holtzhausen and Zerfass 2013, p. 284).

The following characteristics are typical of strategic communication:

- Deliberative and intentional actions
- Directed toward a specific result
- Inclusion of one or more experts who communicate on behalf of an organisation
- Communication with the public (Holtzhausen and Zerfass 2013, p. 285)

These characteristics should render it possible to pool supportive communication measures under a defined strategy and thus generate synergies. This approach calls for integrated communication which is aimed at 'creating a unity of the differentiated sources of the organization's internal as well as external communication in order to convey a consistent image of or reference object for the organization which is suitable for the target group' (Bruhn et al. 2006, p. 17). According to NATO, StratCom refers to the harmonised and appropriate use of communicative strategies and the respective communication skills by its members. This includes, among others, the use of diplomatic means in public and military affairs, but also information operations and psychological operations, depending on what, based on assessment, is considered the best option in order to support the policy of the allies, their operations and activities as well as to promote and adequately convey the aims stated by NATO (NATO 2009a).

The purpose of this approach is to foster awareness for NATO operations and to establish an understanding of and support for them. NATO's communication is therefore complementary to the communication of the allies on a national level. The media should be used to inform the public accurately and proactively about NATO missions. The information activities encompass all methods and measures used in order to directly and indirectly influence the behaviour and attitude of the opposing forces as well as foreign and domestic civilian populations within the context of the military operations. The following activities can be distinguished:

- Informing the domestic population
- Informing the domestic troops
- Informing and manipulating the foreign population
- Manipulating the foreign troops
- Contesting for the public opinion worldwide via the media

Informing the home population and troops is supposed to be limited exclusively to objective, truthful and up-to-date information. The aim of StratCom is to develop a coherent communication strategy that covers all five areas.

A closer look at Germany and France will show how the patterns of civil-military relations shaped the outcome of StratCom in these two NATO member states.

Germany

By means of the Multinational Experiment Series, NATO – with the involvement of Germany, France and the UK – brought forward a concept of StratCom within NATO (All Partners Network 2017; Public Intelligence 2010).

NATO's StratCom aims for close collaboration between civilian and military agencies within the framework of the concept. To the present day, however, Germany has not implemented any such concept. This can be ascribed to historical as well as social reasons. The definition of propaganda, which is often equated with Nationalist Socialist methods and the doctrines of the Third Reich, as well as the definition of defense need to be called into question and redefined. New types of threats and dangers call for continuous adaptation in order to react quickly, proactively, effectively and successfully in a rapidly changing globalised environment (Mölling 2011, p. 1).

Germany agreed in principle to NATO's StratCom, which must be understood as political and subordinate guidelines, without verifying, however, its compatibility with the national communication concepts in advance. Germany lags behind with respect to the adherence and implementation of these guidelines. There are many reservations, particularly regarding psychological operations, which are supposed to influence, by means of new technologies, the perception, attitude and behaviour of chosen people or groups with respect to the goals of NATO (NATO 2009b, pp. 1–2). Other NATO countries also make use of psychological operations for low intensity conflicts in order to influence opposing actors or the population. In this context, the Federal Constitutional Court of Germany draws a line between illegal propaganda and admissible political advertising.

Within the German Armed Forces, there are already concepts for sub-areas of StratCom, such as the press and public relations of the German Armed Forces (Bundeswehr 2016), operational communication (Dienststelle der Streitkräftebasis 2014) and information operations (Bundesministerium der Verteidigung 2014). In this context, the aim is to eliminate the existing contradictions within these sub-areas and to unify the given structures in a standardised concept of StratCom to achieve doctrinal unity. In order to properly implement NATO StratCom guidelines, it is important to develop a corresponding law which is not limited to the application of the guidelines but also involves clarification as to what is permitted within military public relations. Additionally, the legislative basis needs to be compatible with the principles of the Federal Constitutional Court, which have been defined with respect to the admissibility of the state's public relations activities (NATO 2011, pp. 1–6). However, German military operations in Afghanistan have led to increasing civil-military conflicts over functional expectations as the military publicly criticised a lack of strategic leadership and the German government frequently publicly downgraded military operational expertise in Afghanistan.[2]

France

The French Defence White Paper of 2013 under President François Hollande confirmed the defence and security strategy under Nicolas Sarkozy in completing France's reintegration into the military command structure of NATO which it had left in 1966 under President Charles de Gaulle (République Française 2013; Ghez and Larrabee 2009). In line with this trend of French transatlantic reorientation, the French armed forces in 2008 introduced a joint concept of StratCom called *les opérations militaires d'influence* [military influence operations] (Centre interarmées de concepts, de doctrines et d'expérimentations 2012, p. 19). It rests on French experiences in Afghanistan but also in Libya and focuses on the defence of national but also alliance interests, a novum for French military concepts. The concept of *les operations militaires d'influence* is constructed closely to NATO's StratCom. In contrast, *l'influence en appui aux engagements opérationnels* (Centre interarmées de concepts, de doctrines et d'expérimentations 2012) do not comprise such a shared understanding of StratCom. The military influence operations are permanently operational and aim at strengthening the armed forces' credibility towards any targeted audience. It includes all military and diplomatic instruments (Centre interarmées de concepts, de doctrines et d'expérimentations 2011).

The inclusion of diplomatic instruments shows the remarkable merging of civilian and military efforts in France's communication effort. It becomes even more remarkable when the fact is considered that the Chief of Staff – the *Chef d'État-major des armées* (CEMA) – first formulates the military goals and supersedes civilian efforts in the area of operations (Ministère de la Défense 2016a, 2016b). In the German case, no such civilian subordination would be possible. In addition, coordinative efforts exist between the French Ministry of Defence and other ministries involved regarding the alignment of national and international state interests. French civil-military relations have experienced some degree of rift over experiences in Afghanistan and Mali, but they still remain in balanced order. While the CEMA enjoys a rather powerful position in streamlining command structures for StratCom, the civilian staff also plays a comparatively strong role in military development. The French procurement agency (*Direction Générale de l'Armement*, DGA) is centrally involved in the development process, a role stronger than in any other NATO member country.[3] The mix of balanced civil-military relations and a common understanding of concept development led to a French StratCom effort which closely resembles US blueprints. However, an increasingly stronger civilian position in concept development has taken place over the lessons learned process of recent French military operations.

Conclusion

Since the advent of the internet and particularly since the rise of social media, the states and their armed forces have been struggling to obtain information superiority. News coverage can be received 24/7 and often in real-time. With this, the agenda setting capability over activities and operations desired by the military does not exist anymore. Every person with access to the internet can advertise his opinion without geographical boundaries and lessen the credibility of information carriers. Traditional military superiority has lost its superior role in disseminating information and decision-making as long as the military competes against faster-paced media content.

NATO is best advised in taking prudent action in regard to StratCom of multinational contingency operations. Such communication policy must be transparent. But underestimated challenges lie ahead. For instance, information operations that are separate from target groups will undermine any fruitful and coherent policy. This was found in 2010 when the Wikileaks platform shared a Central Intelligence Agency (CIA) memorandum which stated recommendations on how to influence German and French mass media in order to convince their local population to support an extension of ISAF in Afghanistan (Wikileaks 2010). Furthermore, it stated how best to contain opposition against such an extension (Foreign Policy 2015). In addition, a 2015 Pew Survey showed that in eight NATO countries, only two (the US and Canada) had a majority of responders who supported military action in case Russia attacked a NATO member state. French and German responders predominantly supported a position of restraint of their armed forces (Pew Research Center 2015, p. 5). Existing military hierarchies are no longer of any importance for the distribution of information and decisions if they are defeated in the race for correct and fast information.

With this in mind, the introduction and implementation of StratCom poses several obstacles to the armed forces and their political superiors.

First, an increase in the merging of communication efforts traditionally separated between civilian and military actors, like diplomacy and public affairs, poses problems of legitimacy as it becomes harder for audiences to distinguish military communication from psychological operations.

Second, as the leadership of StratCom rests with the political sphere, civilian actors increasingly take part in doctrine building processes, a traditionally military area of accountability. This further increases opposition in the military leadership against StratCom as the German case shows where a gap between the civilian and the military leadership is becoming manifest.

As studies have shown, political leadership has been needed since the end of the Cold War in order to kick off major changes in large organisations, but the military counterpart can also play its part, hampering and therefore shaping singular outcomes of the development process. The process gets more complex when the interaction of civilians and military stakeholders is based on vague procedural norms of formal and informal civil-military relations.

The same goes for military bureaucratic processes. At the core of military organisations is a tendency to stabilise and routinise procedures. On the downturn, they usually resist change of attuned procedures and even more so if these changes result in a decreasing capability for the bureaucracy to acquire resources. The development of StratCom in NATO's member countries has ultimately led to a twofold civil-military conflict. One focuses on increasing frictions in civil-military relations within every NATO member country, as Germany has shown. These frictions ultimately have led to different military concepts in contrast to what was originally intended under military transformation.

The other side of this conflict sheds light on frictions between member countries and NATO's HQs. Over the course of ISAF, frictions among NATO member states have emerged over the use of force in Afghanistan starting in 2006, the number of civilian and military resources during the strategic change in 2009, and in the subsequent handing over of security to Afghan authorities and involvement over the next decade. In this context, StratCom still waits for a coherent implementation. As results in many countries thus far suggest, common efforts will remain vague and uncoordinated. The current crisis with Russia underlines this trend.

During the Cold War, NATO states were more or less strategically aligned and therefore operational concepts were not as hotly debated as in contemporary strategic environments. This means that today, the introduction of a new NATO concept will most likely draw the attention of only some NATO allies while most will accept the necessity of a concept but will not implement it. A case in point is the growing number of so-called NATO Centres of Excellence (CoE). These think tanks are established by a framework nation and accredited by one of two strategic commands of NATO, Allied Command Transformation. However, they only partially report within the NATO command structure. They develop concepts and doctrines but cannot claim authority over concept-building for all NATO members. Instead, they remain dependent on the support of participating countries.

In 2014, NATO accredited a CoE for StratCom based in Latvia. It aims at evaluating, coordinating and assisting in NATO StratCom efforts during exercises and operations. For example, the Centre participated in NATO StratCom assistance to the Ukrainian government in countering Russian media efforts. But so far, only very few member countries participate in exercises and provide financial support.

In addition, there is the outer enemy, the adversary that threatens the security of the Alliance, but NATO countries are fighting far more frequent battles against the inner enemy, that is, the bureaucratic processes of member states' organisation and the political hampering within NATO and member countries. As NATO seems to be pressed to do collective defence and crisis management simultaneously, the challenges become more and more demanding. As such, differing strategic goals and self-willed bureaucratic outcomes of military organisations are hampering the fulfillment of today's strategic and operational goals.

The transformation process was started with the aim to reduce the capability gap between the US and the rest of NATO (Farrell and Rynning 2010). This gap has somehow shrunk even if it still exists. However, a new gap has arisen between those countries that managed to keep moving along the transformation track amid greater opposition at home and those that bowed under this pressure. This places a risk at the centre of the alliance of strategic implications. The twofold civil-military conflict is yet to materialise for NATO and its member countries in light of recent crises in Ukraine, the Middle East, and North Africa. Without balanced civil-military relations these crises will reinforce NATO's internal challenges.

Notes

1   Following the development of the US defence doctrine 'Flexible Response' under President John F. Kennedy in 1961, the Soviet armed forces did not react until the Chairman of the CPSU, Nikita Khrushchev, was dismissed and new ideas in the form of newly appointed officers entered doctrinal development, see Zisk 1993, p. 80.
2   Noetzel 2010; authors' interview with German officers, Berlin, 12 May 2015, Bonn, 26 May 2015 and Stans, 21 July 2015.
3   Farrell et al. 2013, p. 246; authors' interview with French officers, Geneva, 30 November 2015 and Lille, 15 January 2016.

References

Adamsky, D. and Bjerga, K. 2012, 'Military Innovation between Anticipation and Adaptation', In D. Adamsky and K. Bjerga (eds), *Contemporary Military Innovation*. London: Routledge, 188–193.

All Partners Network. 2017, Multinational Experiment (Public Site), viewed 25 April 2017, https://wss.apan.org/s/MEpub/default.aspx.

Avant, D. 1993, 'The institutional sources of military doctrine: Hegemons in peripheral wars', *International Studies Quarterly*, 37(4): 409–430.

Avant, D. 2007, 'Political Institutions and Military Effectiveness', In R. A. Brooks and E. A. Stanley (eds), *Creating Military Power. The Sources of Military Effectiveness*. Stanford, CA: Stanford University Press, 80–105.

BBC. 2013, Q & A: Afghan Taliban open Doha office, viewed 30 June 2016, www.bbc.com/news/world-asia-22957827.

Bruhn, M., Ahlers, G. M., Bobolik, B., Eichen, F., Frommeyer, A. 2006, *Integrierte Kommunikation in den deutschsprachigen Ländern: Bestandesaufnahme in Deutschland, Österreich und der Schweiz*. Wiesbaden: Westdeutscher Verlag.

Bundesministerium der Verteidigung. 2014, Vernetzte Operationsführung, viewed 30 June 2016, www.bmvg.de/portal/a/bmvg/!ut/p/c4/bY2xDoJAEET_6I4zNtp JaIyN2ig2ZIHl2HDckWXBxo93KeycTaaYN5m1L6sXYSUPQilCsE9bNnSs 36YeV29manrkHknmKQUSGgxEj3USNIzQIId_Kvsq4qIIlk4YfKVrHWBP ftAWRvvYvrZomhRRNtdQSN0zSGIzJZawkYVZiaHWlpkr8sxlP7nP4XIrd len-Tm_22kcT1-p1Muj/.

Bundeswehr. 2016, 'Pressesprecher des Einsatzführungskommandos der Bundeswehr', viewed 30 June 2016, www.einsatz.bundeswehr.de/portal/a/einsatzbw/!ut/p/ c4/04_SB8K8xLLM9MSSzPy8xBz9CP3I5EyrpHK9pPKU1PjUzLzixJIqILegK-LW4OFW_INtREQCQiPrH/.

Burk, J. 1998, 'The logic of crisis and Civil-Military relations theory: A comment on Desch, Feaver, and Dauber', *Armed Forces & Society*, 24(3): 455–462.

Burk, J. 2002, 'Theories of democratic Civil-Military relations', *Armed Forces & Society*, 29(1): 7–29.

Centre interarmées de concepts, de doctrines et d'expérimentations (CICDE). 2011, *Assistance Militaire Opérationnelle à une Force étrangère*. Doctrine interarmées DIA-3.4.5.1_AMO(2011), Paris: Ministère de la Défense.

Centre interarmées de concepts, de doctrines et d'expérimentations (CICDE). 2012, *L'influence en appui aux Engagements Opérationnels*. Réflexion doctrinale interarmées RDIA-2012/008_INFLUENCE(2012), Paris: Ministère de la Défense.

Deni, J. 2016, 'Shifting locus of governance? The case of NATO's connected forces initiative', *European Security*. 25(2): 181–196.

Department of Defense. 2007, DoD Strategic Communication Plan for Afghanistan, viewed 30 June 2016, https://docs.google.com/viewer?a=v&pid=sites&srcid=ZGV mYXVsdGRvbWFpbnxhZmdoYW5wb2xpY3lzaXRlfGd4OjI1NjQ1ZjA5OTU3U3 MmFmZWQ.

Desch, M. 1998, 'Soldiers, states, and structures: The end of the cold war and weakening U.S. civilian control', *Armed Forces & Society*, 24(3): 389–406.

Dienststelle der Streitkräftebasis. 2014, Zentrum Operative Kommunikation der Bundeswehr, viewed 30 June 2016, www.kommando.streitkraeftebasis.de/ portal/a/kdoskb/!ut/p/c4/04_SB8K8xLLM9MSSzPy8xBz9CP3I5EyrpHK94 uyk-OyUfL2S1KKixNK0dL2q_ILMvLR8_YJsR0UAFYFhwA!!/.

Dyson, T. 2008, 'Convergence and divergence in post-cold war British, French and German military reforms: Between international structure and executive autonomy', *International Security*, 17(4): 1–46.

Dyson, T. 2012, 'Organizing for counterinsurgency: Explaining doctrinal adaptation in Britain and Germany', *Contemporary Security Policy*, 33(1): 27–58.

English, A. 2004, *Understanding Military Culture. A Canadian Perspective*, Montreal: McGill University's Press.

Farrell, T. 2001, 'Transnational norms and military development: Constructing Ireland's Professional Army', *European Journal of International Relations*, 7(1): 63–102.

Farrell, T. 2002, *The Sources of Military Change: Culture, Politics, Technology*, Boulder, CO: Lynne Rienner Publishers.

Farrell, T. and Rynning, S. 2010, 'NATO's transformation gaps: Transatlantic differences and the war in Afghanistan', *Journal of Strategic Studies*, 33(5): 673–699.

Farrell, T. et al. 2013, *Transforming Military Power since the Cold War. Britain, France, and the United States, 1991–2012*, Cambridge: Cambridge University Press.

Feaver, P. 1998, 'Crisis as shirking: An agency theory explanation of the souring of American Civil-Military relations', *Armed Forces & Society*, 24(3), 407–434.

Feaver, P. 1999, 'Civil-Military relations', *Annual Review of Political Science*, 2(1): 211–241.

Feaver, P. 2003, *Armed Servants: Agency, Oversight, and Civil-Military Relations*, Cambridge, MA: Harvard University Press.

Foreign Policy. 2015, Inside the CIA Red Cell. How an Experimental Unit Transformed the Intelligence Community, viewed 30 June 2016, http://foreignpolicy.com/2015/10/30/inside-the-cia-red-cell-micah-zenko-red-team-intelligence/.

Ghez, J. and Larrabee, S. 2009, 'France and NATO', *Survival*, 51(2): 77–90.

Grissom, A. 2007, 'The future of military innovation studies', *Journal of Strategic Studies*, 29(5): 905–934.

Groitl, G. 2015, *Strategischer Wandel und zivil-militärischer Konflikt. Politiker, Generäle und die US-Interventionspolitik von 1989 bis 2013*. Wiesbaden: Springer VS.

Hallahan, K., Holtzhausen, D., van Ruler, B. Vercic, D., Sriramesh, K., 2007, 'Defining strategic communication', *International Journal of Strategic Communication*, 1(1): 3–35.

Herspring, D. 1990, *The Soviet High Command, 1967–1989: Personalities and Politics*. Princeton, NJ: Princeton University Press.

Herspring, D. 1996, *Russian Civil-Military Relations*. Bloomington, IN: Indiana University Press.

Herspring, D. 2011, 'Creating shared responsibility through respect for military culture: The Russian and American cases', *Public Administration Review*, 71(4): 519–529.

Hochuli, A. 2016, 'Der ISAF-Einsatz und die transformation der streitkräfte in Europa', *ASMZ*, 182(4): 46–47.

Høiback, H. 2013, *Understanding Military Doctrine. A Multidisciplinary Approach*. London and New York, NY: Routledge.

Holtzhausen, D.R. and Zerfass, A. 2013, 'Strategic Communication: Pillars and Perspectives of an Alternative Paradigm', In K. Sriramesh, A. Zerfass, J.-N. Kim (eds), *Public Relations and Communication Management. Current Trends and Emerging Topics*. London and New York, NY: Routledge, 283–302.

Huang, R. 2016, 'Rebel diplomacy in civil war', *International Security*, 40(4): 89–126.

Jackson, C. 2016, 'Information is not a weapons system', *Journal of Strategic Studies*, 39(5–6): 820–846.

Janowitz, M. 1964, 'The military in the political development of new nations', *The Western Political Quarterly*, 17(4): 6–10.

Kaplan, F. 2013, *The Insurgents. David Petraeus and the Plot to Change the American Way of War*. New York, NY: Simon & Schuster.

Mayer, S. 2014, *NATO's Post-Cold War Politics. The Changing Provision of Security*. Basingstoke: Palgrave Macmillan.

Ministère de la Défense. 2016a, Actualité, viewed 30 June 2016, www.defense.gouv.fr/ema/le-chef-d-etat-major/actualite.

—. 2016b. État-major des armées, viewed 30 June 2016, www.defense.gouv.fr/ema.

Mölling, C. 2011, *'Für eine sicherheitspolitische Begründung der Bundeswehr. Zehn Punkte für die Reform der Reform'*, Stiftung Wissenschaft und Politik, SWP-Aktuell, no. 20. Berlin: SWP.

NATO. 2009a, Strategic Communication Policy, viewed 30 June 2016, https://publicintelligence.net/nato-stratcom-policy/.

—. 2009b, Strasbourg-Kehl Summit Declaration, viewed 25 April 2017, www.nato.int/cps/en/natohq/news_52837.htm?mode=pressrelease.

—. 2011, NATO Military Policy On Public Affairs, MC 0457/2, viewed 30 June 2016, www.nato.int/ims/docu/mil-pol-pub-affairs-en.pdf.

—. 2017, NATO Common-Funded Budgets & Programmes, viewed on 25 April 2017, www.nato.int/nato_static_fl2014/assets/pdf/pdf_2014_06/20140611_20140601_NATO_common_funded_budgets_2014-2015.pdf.

Noetzel, T. 2010, 'Germany's small war in Afghanistan: Military learning amid Politico-strategic inertia', *Contemporary Security Studies*, 31(3): 486–508.

Perlmutter, A. and LeoGrande, W. 1982, 'The party in uniform: Toward a theory of Civil-Military relations in communist political systems', *The American Political Science Review*, 76(4): 778–789.

Pew Research Center. 2015, NATO Public Opinion: Wary of Russia, Leery of Action on Ukraine, viewed 30 June 2016, www.pewglobal.org/2015/06/10/1-nato-public-opinion-wary-of-russia-leary-of-action-on-ukraine/.

Posen, B. 1984, *The Sources of Military Doctrine. France, Britain, and Germany between the World Wars*. Ithaca, NY: Cornell University Press.

Public Intelligence, 2010, NATO Military Concept for Strategic Communications, viewed 30 June 2016, https://publicintelligence.net/nato-stratcom-concept/.

République Française. 2013, *French White Paper on Defence and National Security—2013*, Paris: Ministère de la Défense.

Rosen, S. 1988, 'New ways of war: Understanding military innovation', *International Security*, 13(1): 134–168.

Strachan, H. 2010, 'Strategy or Alibi? Obama, McChrystal and the operational level of war', *Survival*, 52(5): 157–182.

Terriff, T. 2006, 'Warriors and innovators: Military change and organizational culture in the US marine corps', *Defence Studies*, 6(2): 215–247.

Terriff, T. et al. 2010, *A Transformation Gap? American Innovations and European Military Change*. Stanford, CA: Stanford University Press.

Wiesner, I. 2010, 'NCO in Germany: Still a long way to go', *RUSI Defence Systems*, 13(1): 82–84.

Wiesner, I. 2013, *Importing the American Way of War? Network-Centric Warfare in the UK and Germany*. Baden-Baden: Nomos.

Wikileaks. 2010, CIA Report into Shoring up Afghan War Support in Western Europe, viewed 30 June 2016, https://file.wikileaks.org/file/cia-afghanistan.pdf.

Zisk, K. 1993, *Engaging the Enemy*. Princeton, NJ: Princeton University Press.

3   Violence reconsidered

Towards postmodern warfare

Artur Gruszczak

Violence is a predominant code written in the protocols of modern war. Since the beginning of the 21st century, it has left an ever-increasing footprint in contemporary politics, both in its domestic and international dimensions. The post-Cold War global order, identified with cooperative security, inter-institutional arrangements, information networking and economic interdependencies, from the very beginning of its convoluted and arduous construction has been seriously challenged by state and non-state actors. The new security environment, permeated by complex globalised networks enabling a constant flow and exchange of goods, ideas, and capital by force of connectivity, was not immunised against violence. Ethnic feuds, territorial litigations, civil wars, insurgencies, organised crime and transnational terrorism have epitomised the 'brave new world' of 'good old wars'. However, warfare is no longer synonymous with coercive force, lethal weapons, combat and assault. It has entailed much more sophisticated psychological, social, technological and economic means and methods which have determined the changing face of violence.

The sudden and hasty career of the concept of Revolution in Military Affairs (RMA) in the last decade of the 20th century was a symptom of 'technophilia': a deeply held military mainstream belief that high technology and novel military organisation will definitely be decisive factors in warfare and enable a considerable reduction in the extent of material damage and psychological harm done by warring parties. RMA was interpreted as the 'application of new technologies into a significant number of military systems combine[d] with innovative operational concepts and organizational adaptation in a way that fundamentally alters the character and conduct of conflict' (Krepinevich 1994, p. 30). Network-centric computerised command and control systems, precision munitions, wide-area sensors and signals intelligence were emblems of the new military era. However, it quickly turned out that RMA loses its sting when it comes to new wars, saturated with barbarity, criminality and organised violence. 'New warriors', including 'paramilitary units, local warlords, criminal gangs, police forces, mercenary groups and also regular armies including breakaway units of regular armies' (Kaldor 2001, p. 8) were much more focused on ideologies, identities, cultures and economic interests than military technologies, doctrines and network-centric solutions.

Since the beginning of the 21st century the gap between the 'war as it is' and the 'war as it was supposed to be' has been quickly widening. The twilight of RMA's conceptual utility, criticism of the renewed counterinsurgency doctrine tackling the challenge of contemporary 'small wars', and the search for a new approach which would frame the rapidly changing conflict environment, triggered a number of different opinions, raising concerns about plausibility of the 'new wars' concept and the RMA revolution. As to the latter, its stimulation effect was already on the wane. Critics put forward new concepts that responded better to singularities of conflicts and violence in the new security environment. One of the most suggestive proposals was the concept of hybrid wars. It has recently come back into vogue in the context of the war in the east of Ukraine and some aspects of military tactics adopted by the Islamic State of Iraq and Sham (ISIS) in Iraq and Syria. However, one could have a feeling that the term 'hybrid war' has been too often used without a clear definition and serious reflection and has become a buzzword covering some unconventional, atypical and complex aspects of today's conflicts.

This chapter offers a critical assessment of dominant 21st century concepts of violence and warfare. Special attention is directed to the still-in-vogue theories of asymmetric conflicts and hybrid wars. However popular in a general discourse, they no longer suffice to explore and explain differentiation, selectivity and singularity as main features of contemporary warfare. They underpin those protocols of war which got mediatised and became appealing because of their simplicity, conventional way of thinking, and binary (black-or-white) approach to ethical behaviour. A concept of postmodern warfare is introduced and elaborated in order to highlight singularities of war and conflict in the complex global-networked environment. It is put forward as an analytical tool useful for analysis and interpretation of contemporary aspects of violence and warfare. An ancillary concept of post-asymmetric warfare is also formulated as a conceptual link between the conventional late 20th century doctrinal thinking and today's unorthodox approach to the nature of warfare in the 'second digital age' (Rid 2007).

Hybrid warfare – a conceptual chimera

'Hybrid warfare' has been in vogue for a decade. Since the series of articles and a monograph published by a retired US Army Lt Col Frank G. Hoffman in the mid-2000s, hybrid war enjoyed a revival in popularity in the aftermath of Russia's aggressive actions in Ukraine in 2014 and escalation of irregular warfare in several North African, Middle Eastern and South Asian countries standing on the brink of chaos and disarray.

The introduction of the hybrid war concept to the strategic discourse was therefore another attempt to formulate an epistemological interpretation of new wars in which the leading powers of the West, with the United States at the forefront, could not capitalise on global hegemony and emerge victorious in the confrontation with a much weaker opponent (Freier 2009, p. 82; Marks 2005, pp. 171–173). Hybridity of armed conflicts was seen as a variable explaining the presence of asymmetry in methods, means and results of military operations. At the same time, it constituted an important contribution to the discussion on American strategic culture in which hegemonic presumption mingled with the powerlessness against the 'silent enemy' hiding behind the curtain of multiculturalism and religious revival.

In the cacophony of voices recommending conceptual frames, theoretical proposals and practical observations on hybridity of contemporary conflicts, the original thoughts on the special character of hybrid warfare somehow vanished, deadened by simplified and overgeneralised statements, such as those defining hybrid war as a mixture of conventional and unconventional tactics simultaneously exhibiting elements of both (Murray and Mansoor 2012). The very notion of hybridity and hybridisation got lost in the rush of interpertations, explanations and insights into the changing face of war in the early 21st century. It is then necessary to take a closer look at the essential meaning of hybridity.

Hybridisation of the contemporary world implies a natural inclination to paradoxes and singularities which not only complicates the cognitive proceeding and hinders an inquiry into the essence of reality, but also creates the need to have various 'tool boxes' at hand, to use research tools of considerable variety, utility and feasibility. Hybridity refers, in its most basic meaning, to mix or – to be more precise – to crossover or interchange structural elements containing systemic features proper for distinct objects, species or beings. Hybridity is a property resulting from the crossing or mixing of characteristic elements belonging to different, often structurally and genetically distinct, distant, opposing objects, organisms or states. Hybridisation is therefore essentially a merging of different features around a single, separate entity, while preserving critical species-specific properties deciding about the 'superiority' of a new hybrid organism in terms of disease resistance, strength and greater adaptability.

In terms of security studies, hybridity is meant as a kind of co-adaptation and convergence of functional elements linking two different systems. Convergence is a systemic approximation that enables progressive, gradual and deliberate bridge-building between separate entities with a view at establishing a universal pattern of 'organizational diversity' (Bonnicksen 2009, pp. 59–68). With regard to security policies, convergence is the integration of physical, logical, informative and personal resources to improve operational efficiency of those responsible for an effective threat prevention and risk management. What is more important, hybridity directly reveals the current features of national and international security consisting in growing complexity, fragmentation and transnationalisation of actors and structures. Human, political and technological networking brings about multi-level connections and relationships which link persons, private entities, companies, state institutions and public authorities having to do with security or insecurity. Hybrid wars demonstrate the complex assemblage of diverse elements of national and international security. This compound system of hybrid security directly reflects hybridisation of risks: political failure, military defeat, financial burden, cultural gaps and psychological shocks are equally relevant to the major players.

The concept of hybridisation is also important because it is goal-oriented, seeking to contribute to a model of security converging diagnosis and analysis of identifiable components of 'security as a practice' with anticipation of threats and projection of 'security as an ideal'. In this context, one can notice since the beginning of the 21st century an important evolution of strategic thinking from network-centric to human-centric and population-centric approaches. It is the soldier and societal environment in which he/she operates that makes the battlefield and predetermines the tools of warfare. The human dimension of military operations is of growing importance given that it is more and more often the decisive element in achieving strategic overmatch and winning the military confrontation. The human dimension involves every participant in warfare, be it soldier, insurgent, contractor, civilian, politician or diplomat. Physical, cognitive, cultural and moral factors are equally relevant for every category of actors involved in an armed conflict.

It has been mentioned in the hitherto scholarship (Mattis and Hoffman 2005; Hoffmann 2007; McCuen 2008; Lasica 2009; Freier 2009) that hybrid war is a combination, synthesis or mixture of structural elements containing systemic features proper for distinct actors, objects or forms of warfare in terms of action, organisation, command, strategy and ideas. In Hoffmann's original view

> hybrid wars blend the lethality of state conflict with the fanatical and pro-tracted fervor of irregular warfare. In such conflicts, future adversaries (states, state-sponsored groups, or self-funded actors) will exploit access to modern military capabilities, including encrypted command systems, man-portable air-to-surface missiles, and other modern lethal systems, as well as promote protracted insurgencies that employ ambushes, improvised explosive devices (IEDs), and coercive assassinations. This could include states blending high-tech capabilities such as anti-satellite weapons with terrorism and cyber warfare directed against financial targets (Hoffman 2009, p. 37).

The amalgamate of different means, tools, methods and areas of violence engenders hybrid warfare as a cross-referential, multi-sectoral process containing elements of each generation of armed conflict, binding them on the battlefield as well as blending certain systemic ingredients at the level of political, socio-economic and virtual (cyber-spatial) interactions. Such intersection refers to the civilisational paradigms, framing the structure of the states, societies and communities. It also addresses the formative processes of cultural and psychosocial origins, as well as collective identities, underpinning the given paradigm. Thus, hybrid warfare reflects the nature of human civilisation in its three main formative stages: pre-modern, modern and postmodern (Cooper 2003), epitomised by barbarism, the power of the nation-state and 'sensitivity' of the technology-driven global structure. In Michael Evans's picturesque description, '[...] we are confronted with a strange mixture of premodern and postmodern conflict—a world of asymmetric and ethnopolitical warfare—in which machetes and Microsoft merge, and apocalyptic millenarians wearing Reeboks and Raybans dream of acquiring weapons of mass destruction' (Evans 2003, p. 137).

Hybrid wars are 'actor-centred', they have increasingly sought to institutionalise inter-personal and group conflicts in diverse societies, combining structural and organisational traits of pre-modern communities (clans, tribes, ethnic groups) and the postmodern hyper-connected world (network structures interlinked by communication technologies and interdependent by the logic of globalisation) (Nemeth 2002, pp. 2–3).

In hybrid conflicts, the organisation of the warring parties is decentralised and partly heterarchical. Well-organised, trained and armed squads under a unified command are at the core. They employ organisational and technological solutions with a high degree of proficiency. Irregular combat groups using guerrilla tactics are in the inner circle. They take advantage of the command, communication and control capabilities developed by the core forces, but they conduct autonomous kinetic and non-kinetic military operations. The outer circle contains supporters and proxies of the warring parties, largely the civilian population who provides support for both the core forces and irregular fighters, although it displays a changing loyalty and the instrumental treatment of its ally.

In hybrid wars, the problem of legitimacy is secondary. Hybrid actors do not necessarily need formal, legal, moral, or ideological legitimisation of their activities. This stands in stark contrast to asymmetric wars, where the warring parties look for different sources of legitimisation of their armed, violent, often cruel actions. International law, moral values, religious truths, democratic elections, financial profits interchangeably motivated reasons behind resorting to force, violence and crime. In hybrid wars, the issue of formal legitimisation is 'diluted' by contending actors seeking to impose their particular reasons and motives, underpinning their basic interests and goals. Quite often their key objectives are material, seeking financial and economic profits and gains. As a result, hybrid warfare contains a strong criminal component (Evans 2003; Bauer 2014).

The wars in Afghanistan, Ukraine and Syria clearly demonstrate the complex assemblage of hybrid elements of national and international security. This complex system of hybrid security directly reflects hybridisation of risks: political failure, military defeat, financial burden, cultural gaps, humanitarian disasters and psychological shocks are equally relevant to the major writers of the protocol of hybrid war.

Post-asymmetric warfare – rebalancing the battlefields

Hybrid wars tend toward the reduction of asymmetry between the actors in the conflict and means and modes of warfare. A classical asymmetric war is a conflict between belligerents with a substantial difference of relative power in which the weaker party uses unconventional methods to undermine an opponent's advantages. In hybrid wars the distinguishing feature is 'convergence of the physical and psychological, the kinetic and nonkinetic, and combatants and noncombatants [...] military force and the interagency community, of states and nonstate actors, and of the capabilities they are armed with' (Hoffman 2009, p. 34). Hybrid wars are a combination of symmetric and asymmetric full spectrum operations with both physical and conceptual dimensions (McCuen 2008, p. 108). In late hybrid wars, drone strikes by Hezbollah at Israel's military targets in strategic terms were a kind of asymmetric action whereas in the technological dimension it was evidently symmetrical move intended to balance Western preeminence. The result was convergence of tactical levels of both adversaries.

The concept of post-asymmetric warfare departs from the assumption that asymmetric conflicts do not exist in practice. The logic of asymmetric action demands an adequate response and brings about assimilation of methods, tools, means and resources engaged in a confrontation. Moreover, it counterposes technological predominance to aboriginal cunning, material wealth to poverty of the underdeveloped, knowledge to ignorance and primitivity. Following Whitehead and Finnström, we agree that:

> The asymmetry of today's warfare [...] is to be located in the very process of bracketing off the allegedly modern from the allegedly premodern or even primitive, by which dominant groups readily ascribe themselves the role of the modern civilizer and defender of democracy, while at the same time identifying the enemy as primitive, somehow less human, or anyway in need of education (Whitehead and Finnström 2013, p. 3).

Asymmetry enforces rebalancing, it makes the weaker seek ways and means of countering advantage of the dominant actor through specific, unique and unpredictable decisions and actions.

Therefore, post-asymmetric warfare heralds the twilight of asymmetric irregular war due to the rebalancing of the adversaries. It is founded on differentiation, selectivity and singularity. It implies institutional singularity yet functional connectivity and operational excellency. Reduction of asymmetry occurs through dynamic interactions based on connectivity as a property of distinct objects. It takes place in those areas which offer most favourable conditions, like cyberspace, information networks, global communication or social interactions.

In military terms, post-asymmetry entails mimetism when it comes to planning and conduct of operations, equipment and training, strategic communication and situational awareness. Mimetism can also be noticed in systemic organisation of the confronting actors. The division into the state and non-state or sub-state actors is still a prerequisite at the outburst of conflict. Post-asymmetric warfare implies the appearance in the arena of confrontation of agile, self-organising heterarchical forces aiming at the reproduction of elements of tactical and operational art intercepted from the state. The state as a set of empowered bureaucratic institutions is still the system of reference for non-state actors. As a result, a tendency towards bringing the state as a critical actor back in the space of strategic rivalry and military confrontation is clearly visible. For sub-state groups violence is closely linked to the state as an organised apparatus claiming monopoly to hold political power over territory and population. Such a Weberian interpretation of the rationale behind violence in political relations and public affairs reinforces the need of state-like forms of organisation of coercion and violence.

If the 2006 Lebanon war was an example of hybrid conflict, the terrorist attack in Mumbai in November 2008 may be taken as the first case of post-asymmetric warfare. The latter case points to the tactics of mimicry and strategic surprise rebalancing potential asymmetry in terms of human and financial resources. A terrorist hit squad, supposedly formed and prepared by Lashkar-e-Taiba (LeT), a Pakistani terrorist group, revealed many important features typical for the state's military organisation, particularly special operation forces, as well as police anti-terrorist units. First, it had gone through a long-term, detailed and diversified preparation phase including reconnaissance, surveillance, operational planning, provision of arms and explosives, tactical training, logistical coordination and target acquisition. The physical preparations included training in marine and urban warfare, use of explosives and weapons and physical fitness. Also, certain skills were systematically developed, especially in the areas of communication, situational analysis and map reading (Acharya and Marwah 2011, pp. 8–9). Intelligence, reconnaissance and surveillance (ISR) were particularly developed and relatively sophisticated. The assault group carried out extensive preoperational surveillance using human reconnaissance, available imagery intelligence and Google Earth pictures. It was organised in a network of dispersed tactical small units with a high degree of operability and mobility (Azad and Gupta, 2011). As a result, at the time of attack, multiple teams attacked several locations at once using guerrilla warfare tactics in an urban environment which combined 'armed assaults, carjackings, drive-by shootings, prefabricated IEDs, targeted killings (policemen and selected foreigners), building takeovers, and barricade and hostage situations' (Rabasa et al 2009, p. 5). Technology played an extraordinary role in preparing, coordinating and enforcing the plan of attack. Aside from the mentioned tactical preparedness, communication devices, like secure cell phones (Blackberries), satellite phones and GPS navigation devices, were massively used during the 60-hour stand-off in Mumbai.

The assailants were organised, equipped and commanded as if they were professional soldiers belonging in special forces of a leading world power. In the first phase of the operation their advantage over local police and military forces was unquestionable. In this sense, they were more effective and better organised than relevant state services. Such a reversed asymmetry signalled the competitive advantage of a post-asymmetric approach, bringing about high strategic gains with relatively scarce material resources invested by the assailants. Although Indian security forces, despite multiple errors and failures committed before and during the terrorist attack, eliminated the hit squad, violence and intimidation focused on the civilian population delegitimised the defence and security functions of the state apparatus for a long time.

The Mumbai case as well as other examples (post-2008 Iraq, Libya in the years 2011–2014) show that one of the most notable features of post-asymmetric warfare is the coexistence of two major dimensions of the conflict:

- Territorial, relating to the classically understood nation-state and traditional ethnic, clan or tribal communities permanently residing in their indigenous territory
- Virtual, having a supra-territorial, cross-border, networked structure that enables communication within the network, the global promotion of the values, ideals and principles, as well as maintenance of reproduction mechanisms.

In terms of territory, war aims to establish and maintain jurisdiction and administrative control in the area, protect the borders delimiting the scope of the jurisdiction, enforce the constitutional and legal norms on the population concerned and ensure the centralised management of natural resources and economic activity.

In the virtual dimension, war redefines the parameters of conflict, and even eliminates some variables, such as territory, natural resources, military organisation and public order. A 'pseudo-state' (Kilcullen 2010, p. 200) may emerge in the virtual realm. It is devoid of the traditional elements of state power, international legal personality and hierarchical organisation, yet it still holds effective tools and methods enabling to shape the international environment, exert impact on the population, multiply financial resources and wage large-scale information campaigns.

When it comes to ethical aspects of post-asymmetry, a comprehensive, effective and feasible solution is sought to the dilemma of uncertainty and insecurity. Normative restraints vanish from the space of conflict because of the logic behind post-asymmetry mimetism. The treatment of civilian population is a telling example in this regard. When the state is trying to set a clear dividing line between the armed forces and civilians, non-state actors deliberately blur this line.

Postmodern warfare – violence in the multi-layered environment

The concept of postmodern war appeared already in the reflection on the nature of contemporary conflict, warfare, and violence. It responded to the postmodernist interpretation of post-Cold War security environment and the speed of globalisation and technological change. The inspiring contributions by Chris Hables Gray (1997), Martin van Creveld (2000) and Christopher Coker (1998, 2014, 2015) reflected the limitations of the debate on violence and war dominated by the technology-driven RMA. Valuable as they are, Gray's, van Creveld's and Coker's ideas only partially correspond to the concept presented in this chapter. I assume that postmodern war is a product of the counter-evolution of warfare since the beginning of the 21st century. It is not a genuine brand-new type of armed conflict; it integrates, blends and reshuffles certain ingredients and features of dominant concepts of 21st century warfare in unconventional way.

The emergence of the postmodern warfare concept has also been determined by the 'de-Westernisation' of strategic studies and military thought. RMA was (almost) dead in the West, but it began its alternative second life in anti-Western areas, especially in the Islamic world. Ralph Peters (2006) called this movement 'counterrevolution in military affairs' while Itai Brun (2010) employed the term 'the other RMA'. The latter observed that, in strategic military terms, a response to an RMA-driven technology-loaded enemy consisted of improving absorption capability, increasing survivability, creating effective deterrence and shifting the war to more convenient though possibly non-conventional areas. Ultimately, the other RMA aimed at winning the war by not losing it, and creating an attrition effect for that purpose (Brun 2010, p. 561).

Postmodern warfare is one of the negative side effects of the rising head-on confrontation between forces of globalisation and technology-driven modernisation and the communities exhibiting structural patterns deeply embedded in traditional cultural, economic, and social relations. It is, in general terms, the 'war on modernity' (Mazarr 2007) that drives the growing wave of discontent, radicalisation and contestation provoking, in its most extreme aspects, violence and terror. Modern warfare, characterised by massive concentration of soldiers and armaments on battlefields, fear of the potential use of unconventional weapons, and gruelling technological race, passed away.

A continuum of the pre-modern to postmodern world is obsolete, the journey through subsequent generations of warfare alike. Rather, the blending of the 'techno-modern' and 'magico-primitive' (Whitehead and Finnström 2013, p. 6) frames a bizarre, obscure, and paradoxical picture of today's war. In the course of the development of conflict and warfare we can observe loops, feedbacks and surprising turnabouts. Discontinuity and counter-evolution underpin the logic of confrontation and use of force. Apparently, warfare does not necessarily involve unconstrained violence resulting from regular military operations, insurgencies and domestic strife. It rather rests on subversion, delegitimisation of legal authority, information operations (mainly in social media), indoctrination and illicit businesses. Technologies are secondary when it comes to the way of engaging in warfare (strategic culture) but they are absolutely crucial as drivers of organised violence and protracted conflicts. A microchip is the essential component of the system of postmodern warfare.

From a strategic perspective, the logic of warfare is fuzzed by overlapping elements belonging to different, often fundamentally disjunctive, spheres of human civilisation framed in substantially distinct strategic environments. Building on Thacker's observation (Thacker 2004) and Chiang's insight into layering as a central tenet of network coordination and resource allocation (Chiang et al 2006), the below approach to postmodern warfare partially integrates elements inherent in other 21st century concepts of war. The most striking aspect of postmodern war is its multi-layered structure, the superposition of distinct, often contrasting, variables corresponding to temporal, organisational and ideational dimensions of human civilisation. Ideology and religion are layered onto political decision making. Ethical issues are layered onto public discourse. Moral principles are layered onto technological advances. Market rules underpin political choices, ethical attitudes and strategic communication. Intelligence and surveillance permeate, overwhelmingly, the organisational structure of states, societies and communities.

Taking into account the above arguments, the following definition of postmodern warfare is therefore proposed:

> Postmodern warfare is a sequential combination of scattered acts of hostility performed in a multi-layered space of confrontation between interconnected units which engage in information-powered operations undergoing in uncertain strategic environment hosting state and non-state actors focused on network building, capital raising, and intelligence acquisition.

Although the term 'violence' is not present in this definition, it is by no means excluded from the concept of postmodern war. My conceptualisation stems from some prerequisites of the original approach to warfare in postmodernity (communication, networking, marketisation of force, psychological influences, cultural cleavages). Evidently, since the beginning of the 21st century we have seen the rise of collective violence. Tilly's remarks properly address this issue:

> Overall, collective violence rises with the extent that organizations specializing in deployment of coercive means – armies, police forces, coordinated banditry, pirate confederations, mercenary enterprises, protection rackets, and the like – increase in size, geographic scope, resources, and coherence (Tilly 2003, p. 44).

In postmodern war, violence is a goal in itself. It breaks off with the close conventional relation to insurgency, counterinsurgency and terrorism, which was typical for asymmetric and irregular warfare. It ceases to be a means of vicious actions, aiming to sow the seeds of fear and intimidate the population. It consists of disruption of systemic structures, protracted destabilisation, escalation of force and deception. Alternatively, new complex networks are established with the view to maximise the power of connectivity and proliferate specific ideas, beliefs and various forms of capital using advanced information technologies.

Connectivity is the feature of postmodern war yet it is embedded in traditional pre-modern civilizations and ethnic communities. It should be conceived as the ability of objects of different nature to integration through exchange. Everybody is connected. It is really hard to live disconnected, isolated and devoid of external linkages and exchange venues. The latter is particularly important since we exchange a whole lot of varied items: information, knowledge, goods and capital, but also feelings, viruses, favours and ideas. What distinguishes nations, communities, ethnic groups, clans, families, institutions or associations is the utility of networks. There are various forms of social organisation and institutionalisation that make use of different types of networks and operate at different levels. There are various degrees of organisation and various forms of connectivity. It is the networking that makes the difference in the whole world of connected components.

No network emerges without connectivity as a feature of its components and as a capability to link them together and how they relate to each other. Connectivity, however, does not require networking or demand solely the ability to interact. Complex networks emerged as a result of the triple revolution in three areas: social networks, the personalised internet and always-available mobile connectivity (Rainie and Wellman 2012, p. ix). These three components do not coexist in many peripheral areas, unlike links and connections embedded in traditional structures of social organisation as well as sophisticated patterns of organisation and management characteristic of advanced modern states and societies.

In terms of contemporary warfare, connectivity brings about a prolonged change in the nature of conflicts. Security-oriented and military-centered networks have been established and developed for one basic purpose: to gain strategic advantage and operational overmatch through asymmetry in defence budgets, military technology, technical solutions, tempo of action and a combination of 'soft' instruments (psychological operations, human terrain system, civil-military cooperation). However, what has worked at a state-to-state level has not necessarily worked in the state-to-non-state field. The reason is that the model of network-centric warfare, characteristic for RMA, was fairly static. It was built on advanced technologies, sophisticated armaments, large-scale IT systems and enhanced kinetic capabilities making up large combat platforms ready for a confrontation with various categories of adversaries. Real networks are not static, their architecture is based on a hierarchy of self-organized hubs or clusters of heavily connected nodes (Barabasi 2002, p. 221). The dynamics of connectivity makes the contemporary security area change and move towards networking and complexity.

Network building, following Galloway (2004, p. 7), can be identified with protocol as a 'technique for achieving voluntary regulation within a contingent environment'. Networking and connectivity as essential features of postmodern warfare frame protocols of war in its typical multi-layered configuration. Thacker (2004, p. xxi) observes that: 'Layering is a central concept of the regulation of information transfer in the Internet protocols'. In postmodern warfare layering enables communication and interaction between different dimensions of conflicts and hostilities. For instance, the emergence of a flexible worldwide network – a terrorist 'society' (Bauer 2014, p. 61) – resulted in the establishment of the protocol of 'long war'. Stimulated by the experience of US-led military operations in Afghanistan and Iraq, enhanced by post-Arab Spring developments in some MENA (Middle East and North Africa) countries (with special focus on Syria, Libya and Egypt), it has covered different layers of state organisation and human activity, ranging from local communities to religious systems. It consisted of the translation of encoded cultural symbols and values into ideological, material and organisational efforts seeking either to 'degrade and ultimately eliminate the scourge of terrorism' or to 'wage holy war against the infidels'. Nebulous organisations with amorphous structure and fluctuating agency – such as al-Qaeda or ISIS (Da'esh) – welcome a simple message contained in the aforementioned protocols because it justifies forms, methods and means of violence inflicted on their social and political environments. Moreover, they do not change the frames of reference, even if the language of communication alters. For example, the 'Jihad against Jews and Crusaders' proclaimed by Osama bin Laden in 1998 was repeated by Da'esh's propaganda machine as the 'war on the forces of kufr [infidels] [...] the enemies of Allah' (Dabiq 2015, 13, p. 14). Western powers and their political-military complexes tried to rationalise the 'long war' against 'violent extermism and radical jihadism' through a reference to military doctrines, such as the refreshed doctrine of low-intensity warfare or the concept of large-scale irregular warfare (Watts et al. 2015, pp. 33–36).

In postmodern warfare, the protocol of war may not be based on logical reasoning. Religious zeal, ideological orthodoxy, personal greed for money and power, deception and 'post-truth' determine, to the highest degree, the behaviour and action of multiple interconnected actors.

Conclusions

The face of war has undergone a tremendous change from the beginning of the 21st century. Revolution in Military Affairs, depicted by one of outstanding experts in strategic studies as 'seductive but nonsensical and extremely costly illusion' (van Creveld 2008, p. 278), gave way to new concepts of warfare devoid of technological hype and doctrinal orthodoxy. The increased pace of systemic changes and growing complexity of political, social, military and economic structures enforced the writing of new protocols of war, encoding violence into political behaviour, culture traits, media contents and individual mind-sets.

Perhaps, as Tilly argues, collective violence concentrates in large waves (Tilly 2003, p. 11). Therefore, the concept of postmodern warfare as a type of violent confrontation between interconnected units in a multi-layered space wired with information and communication networks, responds to the rising wave of violence in its divergent forms: from terrible indiscriminate slaughter to sophisticated high-tech intrusion and disruption. Pre-modern primitive tools meet postmodern advanced devices. As Whitehead and Finnström (2013, p. 3) suggest, in the contemporary realities of war the modern cannot really be separated from the pre-modern. Thomas Rid (2007) remarks that: 'Today's jihad is fought with the methods and weapons of the 7th century and the 21st century simultaneously'. The postmodern adds a unique flavour to the original composition of methods and means used to wage war.

The protocols of postmodern war are written down not only in ideological projects, but also strategic blueprints or doctrinal principles. They can be identified in the rules of strategic communication, public policies, and commercial activities. The latter are particularly relevant because the market frames the distribution and provision of violence as an illicit service. We already pointed out that crime accompanies violence in a hybrid conflict environment. Carolyn Nordstrom (2004, p. 3) noticed that: 'Wars and illicit economies make strange bedfellows'. Alain Bauer (2014, p. 63) coined the term GANGSTERRORISM for the description of dangerous liaisons between terrorist organisations and criminal gangs. The pillage and illicit traffic of valuable cultural heritage and the selling of looted property on the black market is done behind the veil of 'savage wars' waged in 'shatter zones'. The arming of criminal gangs by the state is justified by the war on terrorism.

The protocols of postmodern war are considerably value-loaded and ideologised. They correspond to ideas, political thought, religious truths and moral arguments. Evidently, they highlight war as a natural, although non-optimal, effect of diversity, present not only in the realm of politics, economics and society, but concentrated at the level of norms, values and beliefs. Therefore, social communication and reciprocity are secondary to religious prayer or ideological zeal. To know the enemy does not necessarily mean to understand him or her. Rather, it implies pressure, coercion and destructive influence. Galloway and Thacker, paraphrasing the famous Abraham Lincoln's dictum, concluded that 'best way to beat an enemy is to become a better enemy'. (Galloway and Thacker 2007, p. 98). But it does not require, as both authors suggest, homogeneity of the enemies and a standarised communication protocol enabling justification of hypertrophy of their competing 'war machines'. The moral argument put forward by Galloway and Thacker is relative, depending on what level of confrontation it addresses.

Postmodern war permeates different layers of human civilisation. It accompanies the world population in everyday activities and is constantly present in mass media and virtual communication networks. It constitutes the basic aspect of the global environment and addresses burning ethical and technological issues without helping to resolve them. Ultimately, it is founded on violence, no matter how mystified it is.

References

Acharya, A. and Marwah, S. 2011, 'Nizam, la Tanzim (System, not Organization): Do Organizations Matter in Terrorism Today? A Study of the November 2008 Mumbai Attacks', *Studies in Conflict & Terrorism*, 34(1): 1–16.

Azad, S. and Gupta, A. 2011, 'A Quantitative Assessment on 26/11 Mumbai Attack using Social Network Analysis', *Journal of Terrorism Research*, 2(2): 4–14.

Barabasi, A.-L. 2002, *Linked. The New Science of Networks*. Cambridge, MA: Perseus Publishing.

Bauer, A. 2014, 'Hybridization of Conflicts', *PRISM*, 4(4): 57–66.

Bonnicksen, A.L. 2009, *Chimeras, Hybrids, and Interspecies Research. Politics and Policymaking*. Washington, DC: Georgetown University Press.

Brun, I. 2010, 'While You're Busy Making Other Plans—The Other RMA', *Journal of Strategic Studies*, 33(4): 535–565.

Chiang, M. et al. 2006, Layering As Optimization Decomposition, viewed 16 May 2016, http://netlab.caltech.edu/publications/Layering-IEEEProc-060402.pdf

Coker, C. 1998, 'Post-modern war', *RUSI Journal*, 143(3): 7–14.

Coker, C. 2014, *Future War*. Cambridge—Malden, MA: Polity Press.

Coker, C. 2015, *The Improbable War. China, The United States and the Continuing Logic of Great Power Conflict*. Oxford—New York: Oxford University Press.

Cooper, R. 2003, *The Breaking of Nations: Order and Chaos in the Twenty-First Century*. London: Atlantic Books.

Evans, M. 2003, 'From Kadesh to Kandahar. Military Theory and the Future of War', *Naval War College Review*, 56(3): 132–150.

Freier, N. 2009, 'The Defense Identity Crisis: It's a Hybrid World', *Parameters*, 39(3): 81–94.

Galloway, A.R. 2004, *Protocol: How Control Exists after Decentralization*. Cambridge, MA and London: The MIT Press.

Galloway, A. R. and Thacker, E. 2007, *The exploit: a theory of networks*. Minneapolis – London: University of Minnesota Press.

Gray, C. H. 1997, *Postmodern war: the new politics of conflict*. New York – London: The Guilford Press.

Hoffman, F.G. 2007, *Conflicts in the 21st Century: The Rise of Hybrid Wars*. Arlington, VA: Potomac Institute for Policy Studies.

Hoffman, F. 2009, 'Hybrid Warfare and Challenges', *Joint Force Quarterly*, 52: 34–39.

Kaldor, M. 2001, *New and Old Wars. Organized Violence in a Global Era*. Stanford, CA: Stanford University Press.

Kilcullen, D.J. 2010, *Counterinsurgency*. Oxford and New York, NY: Oxford University Press.

Krepinevich, A.F.Jr. 1994, 'Cavalry to Computer: The Pattern of Military Revolutions', *The National Interest*, 37: 30–42.

Lasica, D.T. 2009, *Strategic Implications of Hybrid War: A Theory of Victory*. Fort Leavenworth, KS: School of Advanced Military Studies United States Army Command and General Staff College.

Marks, T.A. 2005, 'Counterinsurgency and Operational Art', *Low Intensity Conflict & Law Enforcement*, 13(3): 168–211.

Mattis, J.N. and Hoffman, F. 2005, 'Future Warfare: The Rise of Hybrid Wars', *U.S. Naval Institute Proceedings Magazine*, 132(11): 30–32.

Mazarr, M. J. 2007, *Unmodern Men in the Modern World. Radical Islam, Terrorism, and the War on Modernity*. New York: Cambridge University Press.

McCuen, J.J. 2008, 'Hybrid Wars', *Military Review*, 88(2): 107–112.

Murray, W. and Mansoor, P. (eds). 2012, *Hybrid Warfare: Fighting Complex Opponents from the Ancient World to the Present*. New York, NY: Cambridge University Press.

Nemeth, W.J. 2002, *Future War and Chechnya: A Case for Hybrid Warfare*, Monterrey, CA: Naval Postgraduate School.

Nordstrom, C. 2004, *Shadows of War: Violence, Power, and International Profiteering in the Twenty-First Century*. Berkeley and Los Angeles, CA: University of California Press.

Peters, R. 2006, 'The Counterrevolution in Military Affairs', *The Weekly Standard*, 6 February, viewed 11 May 2016, www.weeklystandard.com/print/the-counterrevolution-brin-military-affairs/article/7835.

Rabasa, A. et al. 2009, 'The Lessons of Mumbai', *RAND Occasional Paper* 249. Santa Monica, CA: RAND Corporation.

Rainie, L. and Wellman, B. 2012, *Networked. The New Social Operating System*. Cambridge, MA and London: The MIT Press.

Rid, T. 2007, 'War 2.0', *Policy Review*, web special, viewed 6 April 2010, www.hoover.org/research/war-20.

Thacker, E. 2004, 'Foreword: Protocol Is as Protocol Does', In A.R. Galloway (ed), *Protocol: How Control Exists after Decentralization*. Cambridge, MA and London: The MIT Press.

Tilly, C. 2003, *The Politics of Collective Violence*. Cambridge: Cambridge University Press.

Van Creveld, M. 2000, 'Technology and War II: Postmodern War?', In Ch. Townshend (ed), *The Oxford History of Modern War*. Oxford and New York, NY: Oxford University Press.

Van Creveld, M. 2008, *The Changing Face of War: Combat from the Marne to Iraq*. New York, NY: Presidio Press.

Watts, S., Polich, J.M. and Eaton, D. 2015, 'Rapid Regeneration of Irregular Warfare Capacity', *Joint Force Quarterly*, 78: 33–36.

Whitehead, N.L. and Finnström, S. 2013, 'Introduction: Virtual War and Magical Death', In N.L. Whitehead and S. Finnström (eds), *Virtual War and Magical Death: Technologies and Imaginaries for Terror and Killing*. Durham and London: Duke University Press.

4 Private security, military companies and foreign fighters

Possible interactions and potential practical implications

Iveta Hlouchova

Introduction

The phenomena of both private security actors and foreign fighters have attracted attention in academia, political circles and the media, as they have increasingly become defining features of the 21st century security environment. Given that private security actors and foreign fighters usually operate in the same theatre of conflict (or areas diversely related to various conflict zones) and are neither isolated, nor immune, from the activities of one another, various interactions in behaviour and actions between the two phenomena can be traced and identified.

As both the private security actors and foreign fighters gain more significance in the practical dimensions of contemporary conflict, it is necessary to explore potential interactions. These interactions can be positive as well as negative in their nature. We must try to identify the most important practical implications these interactions can have for efforts aimed at tackling, countering and eliminating the trend of increasing numbers of volunteer foreign fighters taking up arms in armed conflicts.

These interactions can be of a varied nature and unfold at different levels. They can be distinguished by diverse links towards state representatives and can stem from various ideological reasons. Still, the understanding of the relationship is necessary to counter the rising trend of foreign fighters with all the related challenges and problems at hand. Addressing the variation of this relationship can also help establish a more effective legal framework in terms of regulations and oversight for the private security market in various geographical areas.

The aim of this chapter is to bring attention to this issue and provide a primary pilot research platform that can be utilised in future research of this topic and identification potential practical implications in this regard.

Conceptual distinctions

With the end of the Cold War, a new security environment has evolved into its contemporary character, distinguishable by a number of significant changes (especially in terms of the state's monopoly over the legitimate use of force, which has been gradually eroding in various directions, seen further in this chapter). New actors, as well as tools and procedures these actors apply in their activities and operations across a broad range of possibilities, have emerged in the new post-Cold War security environment. One of the major trends is the process of privatisation of security that has produced a variety of different actors and stakeholders operating in and structuring the current security environment. In the broadest terms, privatisation of security manifested itself in the wide variety of non-state actors with different ontological features and motivational, operational, functional and organisational aspects. Generally, two major trajectories can be traced, distinguishable by important similarities as well as fundamental differences, i.e., bottom-up and top-down. Both differences are perceived through the prisms of directions and levels, where the bottom line is drawn on the state level. The bottom-up approach captures all the processes on the transnational level, while the top-down trajectory analogy indicates the existence of security actors on the sub-state, regional and local level. Private security and military companies are widely considered to be the most notorious representative of the bottom-up approach, whilst there is a broad array of paramilitary, insurgent and criminal entities and groupings emerging as result of top-down processes within the overall privatisation of security.

The author of this chapter focuses on the analysis of the potential existing relationships and interactions between private security and military companies and foreign fighters, and the potentially strong practical implications those interactions might have. Both actors have enjoyed increasing significance in the contemporary security environment. They have also become increasingly present and active in the hot spots and conflict zones all over the world.

Before proceeding with the analysis, a conceptual distinction between the two phenomena has to be drawn and clarified for analytical purposes in order to create the initial minimal understanding of the subject matter of this chapter. Furthermore, the theoretical and conceptual framework will be addressed.

Both private security and military companies and foreign fighters are non-state actors in nature, and a number of fundamental similarities in their modi vivendi, as well as modi operandi, can be identified. These include challenges existing with their accountability, attribution and applicability of the Additional Protocol to the Geneva Conventions (1977). Tackling both, mainly the practical terms, substantial overlaps in their activities and operations in particular, continue to shape the security environment and the main characteristics of the two actors, especially in the last few years. However, significant differences in terms of the genealogy, motivation and ideology, strategy, operational activities, applied tactics, techniques and procedures can be identified. Also, their organisational structure and hierarchy, as well as appropriate legal and regulatory framework or their relationships with and links to the official state structures, can also be observed.

Before resorting to conceptual discussions and clarifications of the two central terms in the text of this chapter below, one of the common features of the two studied actors must be emphasised, i.e., to what extent both private security and military companies and foreign fighters must utilise the concept of mercenarism, particularly considering the legal, regulatory and practical implications and consequences.

Under the international humanitarian law, the Additional Protocol I to the Geneva Conventions relating to the Protection of Victims of International Armed Conflicts (1977, article 47) defines a mercenary as a person who:

1 Is specifically recruited locally or abroad in order to fight in an armed conflict
2 Does, in fact, take a direct part in the hostilities
3 Is motivated to take part in the hostilities essentially by the desire for private gain and, in fact, as promised by or on behalf of a Party to the conflict, material compensation substantially in excess of that promised or paid to combatants of similar ranks and functions in the armed forces of that Party
4 Is neither a national of a Party to the conflict, nor the resident of territory controlled by a Party to the conflict
5 Is not a member of the armed forces of a Party to the conflict
6 Has not been sent by a State which is not a Party to the conflict on official duty as a member of its armed forces (United Nations 1977)

Generally, the definition of mercenary is very restrictive in its nature, since it requires that all six conditions must be cumulatively fulfilled. Sufficient evidence to prove the primary motivation by private gain is also required. As derived from the above-mentioned definition of mercenary, potential theoretical and practical overlaps exist between mercenaries and private security and military companies. This can also be seen in mercenaries and foreign fighters, since members of both studied entities can be classified as mercenaries, even when meeting other criteria. Nevertheless, the aforementioned definition also allows for drawing a clearer distinctive line between the respective entities. The dominant feature of the mercenary definition is the criterion of being recruited locally, which mostly goes against the nature of foreign fighters and sometimes against the practices of private security and military companies. Another significant challenge for classifying foreign fighters and private security and military companies as mercenaries is the unclear and unspecified nature of direct participation in the hostilities. Many private security and military companies, and sometimes foreign fighters, specialise their activities and operations in supporting roles and functions, like logistical, advising, assisting or training. The question of where the line of taking direct part in the hostilities lies becomes important. Another problematic aspect of the classification of private security and military companies and foreign fighters as mercenaries is their relations to the armed forces of a party to the conflict. In contemporary conflict zones and on contemporary battlefields, both private security and military companies and foreign fighters operate alongside armed forces of the parties to the conflict. These relations vary in the level of their institutionalisation and autonomy of actions, but are mostly publicly recognised by the conflict parties.

A more detailed conceptual clarification of both actors follows. Importantly, however, the author of this chapter does not seek to provide exhaustive discussion on the nature and major characteristics of the respective security actor, nor does she aim to compare them. The conceptual introduction of the two core concepts is intended to set up the theoretical framework and draw the conceptual line for the purposes of the research and analysis contained in this chapter by presenting and analysing the most significant features of the respective subjects.

Foreign fighters (FFs)

Historically, foreign fighters, sometimes labelled as volunteers, played significant roles in various armed conflicts. The Syrian civil war, which started in 2011, transformed the general character of the foreign fighters' trend into a more sophisticated and systematic form. In this regard, a new sub-trend has emerged within the broader category of foreign fighters, i.e., foreign terrorist fighters. Foreign terrorist fighters are only one of a number of potential sub-categories of the foreign fighters' category, reflecting the predominant feature in terms of their activities, operational methods and techniques, combat tactics of choice as well as organisational classification, know-how and skill set at their disposal.

So far, Thomas Hegghammer developed the most commonly accepted definition of a foreign fighter. For an individual to be classified as a foreign fighter, the following four requirements need to be met: (1) has joined and operates within the confines of an insurgency; (2) lacks citizenship of the conflict state or kinship links to its warring factions; (3) lacks affiliation to an official military organisation; (4) is unpaid (Hegghammer 2011). This definition can be challenged from the practical point of view on all four points. First, foreign fighters do not have to join only an insurgency, even though this assumption appears to be valid in the absolute majority of contemporary cases. However, the validity of the insurgency clause depends on how one defines an insurgency as well as situations where wannabe foreign fighters join a state's armed forces (with no official contract). In the second criterion, Hegghammer was probably dealing with the 'foreign to whom' question (see below). His definition largely limits the number of individuals meeting the definition of a foreign fighter. Nonetheless, in the majority of contemporary cases, foreign fighters possess a dual citizenship, or are second and third generation immigrants with religious or ethnic (thus still kinship) links to the combat zones. Furthermore, the third criterion does not mean that an individual who is a service member cannot decide on his or her own and voluntarily join a conflict party. Hegghammer's criterion here is concerned more with the relation to the state – a foreign fighter is neither sent by a government nor does he or she act under orders from an official military organisation, not even as an operative of a clandestine operation. The evidence also shows that contrary to Hegghammer's fourth criterion, foreign fighters receive money in a form of salary or rewards. The profit motivation, however, is not the primary one, mostly it complements other, more significant motivational drivers.

The fundamental challenge researchers encounter while exploring and analysing the subject at hand is an appropriate, necessary and sufficient answer to the question of whom the fighters are foreign to. The underlying idea behind this conceptual challenge is represented by the fact that many individuals (particularly taking part in the contemporary conflicts) sometimes hold dual citizenship. What then must be addressed is, for instance, the mechanisms of the classification of a Syrian-American as a foreign fighter (for both legal and research purposes), in case this respective individual left the US to join any conflict party in the Syrian civil war.

As already mentioned, foreign fighters are often considered volunteer fighters. The factor of volunteering is indicative of existence of free will of those individuals labelled foreign fighters. The decision to join a conflict party in various geographical areas is his or her own decision, based on free will and personal motivation. Two important variables are significant from both analytical and practical perspectives: (1) the factor of manipulation affecting an individual's free will and decision-making,[1] and (2) the multifaceted variations in the character of conflict zones where individual foreign fighters engage in combat. The choice of a conflict is usually influenced by the amount of knowledge, contacts, availability, resources and means an individual has at his/her disposal, as well as on some of the individual's motivational drivers (for example, when a person wants to bring attention to a certain conflict which in their view is widely overlooked, or significant). Significant practical and analytical challenges are also posed by the unclear drivers that would sufficiently explain the variations in the individual's choice between domestic and foreign fighting (see Hegghammer 2013).

Significant analytical and even more practical implications of the subject matter at hand are also dependent on a proper understanding of not only WHY individuals become foreign fighters, but also HOW. Understanding the process of radicalisation, particularly decision-making and planning, is critically important for establishing an effective framework for countering violent extremism. Given the highly individualised nature of the radicalisation processes, such counter-extremism strategies need to be as narrow in their scope as possible.

Another analytical and practical challenge is represented by a recent trend of individuals voluntarily coming to a conflict zone, not for fighting but supporting a conflict party by various means. This is particularly true in the case of females joining the Islamic State in order to support their project of state-building (the caliphate). This fact also poses an important challenge for any attempts at conceptualisation since these supporters are often labelled as foreign fighters too. Thus, the phenomenon of foreign fighters is not gender specific. On the contrary, the gender proportionality has started to close up recently.

In order to capture the phenomenon in its modern complexity, and in order to set the theoretical and conceptual framework of this paper, the following definition of foreign fighters is more appropriate and, practically, more applicable. Even though the definition is broad, it still captures the dynamics of the phenomenon in a more comprehensive manner than the definition developed by Thomas Hegghammer (see above). Foreign fighters are (1) individuals (militants, supporters) operating in conflict zones outside their home country (where they were born and lived); (2) outside the ranks and structures, or without any active links or official business (including in a clandestine operation) with regular armed forces, integration into their ranks can still have FF qualities; (3) monetary motivation is mostly irrelevant or inferior to other drivers of an ideological, political, religious, ethnic, etc. nature.

In the contemporary security environment, two major trends within the foreign fighters phenomenon can be identified as two polarised flows of individuals pouring into the same conflict zone. These are: (1) Islamist foreign fighters coming to (mainly) Iraq and Syria to join the so-called Islamic State's forces, and (2) the anti-IS foreign fighters' movement, coming to the same conflict zone and joining opposing conflict parties in order to engage in combat against the Islamic State fighters. A new analytically significant variable has crystallised from this most recent dichotomy, i.e., the factor of past combat experience of foreign fighters, given that most of the anti-IS foreign fighters are often veterans of the foreign military engagements in Iraq and Afghanistan. Also, a large number of combat-experienced individuals joined the Islamic State.

Private Security and Military Companies (PSMCs)

The second issue addressed here is the phenomenon of private security and military companies. PSMCs are corporate actors which have been taking an increasingly significant role in the modern warfare. The PSMCs' existence and operations are tied to the post-Cold War security environment where we can notice a rapid boom in privatisation and corporatisation of security. PSMCs are based on entrepreneurship. They represent legal business entities which are legally registered. PSMCs are built along business structures, corporate standards, principles and values.

The performance of PSMCs is based on the mechanisms according to which security goods and services are either financed or delivered or both by an entity other than government.[2] States are the most common clients of PSMCs since governments outsource the monopoly over the legitimate use of force to non-state corporate entities. The relations of PSMCs with state security structures and systems in general stretch from mutually supportive activities (cooperation) to mutually undermining roles and functions (competition, conflict). In certain situations, a duplication of activities can be identified. Theoretically speaking, the two actors can be in a mutually neutral relationship in a certain area, without any significant influences affecting each other. Potentially, the relationship between PSMCs and state structures in an area of operation can be absent. The customer/client base in general stretches beyond the nation states to include various non-governmental organisations, transnational corporations or international organisations.

Given the very specific nature of the services provided by PSMCs, a proper regulatory framework and mechanism is required. To this day, the global private security market remains insufficiently regulated. The most significant framework is embodied in the Montreux Initiative, i.e., the Montreux Document (International Committee of the Red Cross 2008), the Montreux +5 Conference and the Montreux Forum. This joint initiative of the government of Switzerland and the International Committee of the Red Cross does not, however, create a new legal framework, but merely gathers relevant international humanitarian law rules and proposed best practices applicable to PSMCs operating in the conflict and post-conflict environments. Any attempts and efforts to design an effective regulation of a PSMCs' status and activities are challenged by two major factors. These factors are (1) the fact that it is the individual states which are presumed to create an effective regulatory framework of the private security market, even though it is the very same entities, the national governments, that constitute the greatest portion of the PSMCs' customer base, and (2) the fundamentally differing standpoints of the states *from* which PSMCs operate and those *in* which PSMCs operate.

The second fundamental regulatory initiative is represented by the International Code of Conduct (ICoC) for the private security industry and market, which is based on the principle of a voluntary multi-stakeholder mechanism. Its Association (ICoCA) is comprised of representatives from the industry, states and civil society. As the Montreux Document does not create any new binding rules, it may appear that it is the ICoC which constitutes the most significant framework for the regulation of the PSMCs. This can be seen in terms of provisions relevant to licensing, registration, vetting and training of personnel, limitations on the scope of permissible activities and accountability for violations and remedies for victims.

The portfolio of goods and services offered by PSMCs is vast. For research purposes, this portfolio is divided into a number of different classification clusters. The basic dichotomy, with significant practical implications, is the differentiation between private security companies (PSCs) and private military companies (PMCs). Usually, PSCs is a broader term to which the concept of PMCs is subordinated, as it is narrower in the spectrum of goods and services they provide. PMCs are distinguishable by the higher security services (like personal security detail, training, advising, assisting, surveillance, intelligence and the like). On the other hand, PSCs provide less kinetic and more static services, including static guarding, logistical supplies support or consultancy.

Peter W Singer, a prominent private security expert, differentiates between private security providers, consultants, and supporters. He describes the specific portfolio of goods and services of each of the categories on the tip-of-spear analogy, i.e., its geographical and functional proximity to front lines and, potentially, also to the direct participation in the hostilities (Singer 2007). Other potential divisions of the spectrum of PSMCs' services include:

1 Offensive combat functions
2 Military and security expertise
3 Armed services (i.e., guarding)
4 Military support (Tonkin 2011)

Some of the services of contemporary PSMCs cannot, however, be easily put into one single category of PSMCs services. For instance, companies providing services specialising in hostage negotiations or companies with expertise in intelligence gathering. Such types of services produce additional controversies and challenges the very existence and activities of PSMCs and their oversight, control, monitoring and regulation.

Most of the controversies surrounding PSMCs in the contemporary security environment are predominantly linked to the apparent lack of their accountability, where decision-makers, security practitioners and the ordinary public still suffer from what the author of this paper calls the Blackwater syndrome. The pressure that started to manifest as the Blackwater syndrome led many politicians and corporate decision makers to leave the public arena and thus strengthen the potential of PSMCs to operate in politically sensitive security fields on behalf of a government. Most recently, PSMCs provide political representatives of states with the plausible deniability they seek to claim a light footprint strategy in politically sensitive conflict zones.

Possible interactions of PSMCs and FFs

In order to provide a systematic and structured analysis of what interactions can possibly exist between PSMCs and FFs in the contemporary security environment, and what potential practical implications can be drawn from this analysis for combating terrorism strategies, legal frameworks, regulatory initiatives and future military engagements, potential contact points, contact forces and contact mechanisms between the two phenomena need to be identified.

Furthermore, parameters, scope and range of these interactions (at various levels) must be determined. What researchers must look for is whether the interactions, contact points or mechanisms happen to be coincidental, or, whether the PSMCs-FFs interaction is deliberate and intentional. The aspect of active vs passive contact points, contact forces and contact mechanisms is also important to address. The fundamental structure of the analysis of the PSMCs-FFs interaction is the spectrum of cooperation-competition-conflict-duplicity-neutral-absence of their roles and functions in respective security environments, areas and fields.

The *business interoperability* framework appears to be very useful in determining the primary contact points, contact forces and contact mechanisms. The business interoperability framework is commonly applied in the business environment, mostly among individual businesses, but also in the relations between businesses and governments. Efforts to apply this interoperability framework, even in a modified version, to interactions between a business entity and a non-state actor in a security environment, can prove to be relatively challenging but not impossible. Doing so can help to draw necessary and useful conclusions out of the framework's application and the following analysis, mainly given the focus on the potential mutual interoperability, i.e., the ability to work together to deliver services in a seamless, uniform and efficient manner across multiple organisations and systems. Interoperability is viewed as a critical factor in determining the range of the spectrum of possible PSMCs-FFs interactions. The application of a modified business interoperability framework allows researchers to identify any potential functional, organisational, ideological, motivational, strategical, operational, tactical, logical, contextual, conceptual and semantic interoperability.

Adhering to the business interoperability framework structure leads researchers to define possible strategic alignment requirements, consistency and a common vocabulary, as all are required to measure an organisation's performance. It also helps to identify how common objectives (if any) will be realised, what agreed principles are necessary and available, and how the roles of the two phenomena relate to each other, in what manner and to what extent. Specific forms of the possible PSMCs-FFs interactions can significantly enhance or diminish each other's capabilities.

Competition / conflict

Potentially competitive and/or conflicting interactions between PSMCs and FFs are determined by the level of (in)compatibility of interests, goals and objectives of the two actors, and on the stakes that are exposed and can be affected by these interactions. As aforementioned, in the efforts to determine the possible interactions and potential practical implications of the contacts between PSMCs and FFs, the nature of such contact needs to be determined in terms of whether and how much direct or indirect and active or passive the contact is, and what it means in practical terms.

The level to which the possible PSMCs-FFs interactions are direct or indirect and active or passive can be determined by applying Singer's tip-of-spear analogy to both sides of the relationship, mapping the contact points, contact forces and contact mechanisms that are in opposition to each other and whether they are potentially subversive and undermining. A significant variable that substantially affects possible PSMCs-FFs interactions is the factor of the third party's involvement and the level of this involvement (particularly in the situation, where the third party involved is a national state, acting as a customer of a respective PSMC and being a target of a group of FFs at the same time).

Very important variables in the research of possible PSMCs-FFs interactions are the narratives each side uses based on how they perceive each other. Given the specific context and nature of their ontology, existence, functions and activities, there is a great potential for playing the so-called delegitimisation game, i.e., both sides' efforts to strip the other side of their perceived legitimacy in combat, the conflict area and the field, as well as international security, political and economic environment.

When researching the competition/conflict potential in the possible PSMCs-FFs interactions, it is the magnitude, the scope, the type, the frequency and the timing of the contact point, contact forces and contact mechanisms that determines the level and the manifestation of the competition/conflict potential. Potentially, very interesting findings can also be achieved when the factor of the first-mover advantage is included in the analysis, particularly while studying the likelihood of a response in potential action-reaction dynamics.

Importantly, conflict potential must be differentiated from competitive potential in the possible PSMCs-FFs interactions. Conflict is represented by a state of discord case by the actual or perceived opposition of needs, values and interests. Competition is one of a number of subtler forms of conflict, among them, rivalry or struggles for power and favor. The potential for competition between PSMCs and FFs is very limited.

Levels of cooperation

On the other side of the spectrum of possible PSMCs-FFs interactions is the potential for cooperation, including the potential for a merging of dynamics between the two sides, causing each other to modify, adjust and hybridise. The fundamental factor that comes before any potential cooperation is trust, and its level and nature. In the presumed situation, trust is expected to be high and based on personal ties and contacts.

In the general business environment, the level of cooperation reaches various scopes, ranges and volumes, and acquires various forms and characteristics in many types and for various reasons. A prominent business and organised crime researcher Phil Williams (2002) categorises six basic forms of cooperation between two distinguishable entities:

1 Strategic alliance
2 Tactical alliance
3 Deals and reciprocal services
4 Exchange/swap
5 Constant, long-term delivery of services
6 Ad hoc, short-term delivery of services

Considering the original natures of both the PSMCs and FFs phenomena, respectively, the form and level of cooperation that seems to be the most likely to materialise in the practical form is the type of constant, long-term delivery of services, as the PSMCs potentially provide a very useful platform for facilitation of the FFs flow into combat zones. Further research in this regard is, however, strongly urged and required.

Potential practical implications

One of the number of various potential practical implications of the possible interactions between PSMCs and FFs that requires attention is the merging dynamics between the two phenomena at hand. PSMCs are increasingly losing their profit-only-driven character and gradually narrating the importance of values in their operations and activities. These factors have led to a gradual hybridisation of the private security market. Particularly in reaction to the rapid growth of the FFs phenomenon, with all its specific characteristics and tendencies, the trend of privatisation of security in the sense of the PSMCs phenomenon has started to evince indicators of losing certain aspects in its primarily profit-driven and profit-oriented nature, while accentuating certain ideological features and, first and foremost, a number of core values of democracy and respect to human rights. An alternative model of PSMCs has started to emerge, mainly in the Western private security market.

This alternative model has also been influenced by the variable of the rising importance of the financial aspect of the FFs phenomenon, not necessarily in terms of motivation of the FFs, but the importance of obtaining enough resources for travel to the conflict zone, an engagement in the conflict zone and ensuring salaries or rewards. Recent evidence indicates that the FFs phenomenon has acquired a number of business-like features, especially in a sense that it is getting more systematic, sophisticated and organised, with the financial aspect playing an important role. The issue of funding has become an increasingly relevant and important feature of the FFs phenomenon.

The merging dynamics of the two phenomena at hand is obvious. It has already been manifested in the creation of such organisations as the Sons of

Liberty International, the American Mesopotamian Organization and the 1st New Allied Expeditionary Forces.[3] They provide the same services that can be found on the portfolio of security services offered by many PSMCs (like training, logistical and material support, consultancy, intelligence gathering, weaponry, medical assistance, etc.), even though they label themselves as non-profit organisations. The nature of these entities, their structures, the way these entities function, and the services they provide increasingly resemble PSMCs, with the only exception of stressing the ideology and value, not the money, as the main driving force. This argument is supported by the policy of the Kurdish Peshmerga forces, stating that while individual fighters are not always accepted, volunteers working as military trainers share their expertise to support those on the front lines of the fight against the IS forces.

These dynamics pose significant additional challenges to both national and international legal regimes and regulatory frameworks of both the default phenomena of PSMCs and FFs. PSMCs facilitating the FFs flow by various means (provision of training, consultancy, etc.) can face potential legal charges of providing material support for terrorism in the case of an individual, acting in the capacity of a PSMC and joining a conflict party as an FF, chooses a group that is designated as terrorist. Furthermore, the more systematic, business-like nature of the FFs phenomenon poses a challenge of establishing links to transnational organised crime networks.

Moreover, in practical terms, these dynamics can also potentially have significant implications for counter policies and counter strategies, including combating terrorism strategies, as it is the potential PSMCs have in combating terrorism strategies that can be exploited to its highest possible degree, including in terms of countering propaganda that has proven to be one of the predominant motivational drivers of FFs. Another significant practical implication of a proper analysis is the official and unofficial contact the business environment is usually full of that can be, for example, leveraged in situations of hostage negotiations, for intelligence gathering purposes or training of the static guard personnel at the critical infrastructure points and nodes.

Conclusions

This chapter offers a preliminary inquiry into a topic that has a number of significant practical implications to the strategies, doctrines, and policies, as well as legal and regulatory frameworks of various aspects of the contemporary global security environment. Given the limited scope of the paper, the basic framework of the research has been established by creating a background for future studies.

The identification of various contact points, platforms, forces and mechanisms between the activities of PSMCs and those of FFs needs to be based on the proper understanding of the two phenomena, the context in which they operate and the way they function. PSMCs and FFs can function and interact in both mutually supportive and mutually subversive (or undermining) manner, after the contact points, contact forces and contact mechanisms are identified.

More attention should be also given to the slightly merging dynamics of the possible interactions between PSMCs and FFs with all the practical implications it may have. The unsatisfactory regulatory framework of PSMCs, as well as FFs per se, is even further challenged by the recent trends of going more corporate, as the FFs were able to exploit some of the potential PSMCs as security actors have.

The author of this paper specifically stresses the need for exploration and implementation of the potential PSMCs have in the fight against FFs fighting alongside the Islamic State and other terrorist entities or security threats. This potential should be analysed through the prism of the potential contact points, contact forces, and contact mechanisms existing between PSMCs and FFs, as well as the risks and challenges PSMCs (can) pose in the fight against terrorism and FFs in practical terms.

Another critically important issue is the potential of PSMCs acting as facilitators of the FFs flows. Such a facilitation could be direct (in the form of fighters going through PSMCs) or indirect (exploitation and leveraging of many formal as well as informal contacts PSMCs usually have in conflict zones). This aspect and its practical implications potentially affect the counter policies, counter strategies and legal and regulatory provisions of both the phenomena at hand significantly, and deserve further attention.

Notes

1 The potential to be forced to go and join a conflict party in exchange for certain structural concessions represents a significant factor related to the issue of manipulation. Even though these individuals usually do not exhibit their free will and do not make their own decision, they are often still labelled as foreign fighters, thus influencing the statistics as well as variables leading to inconclusive or inaccurate research results and/or practical implications. For instance, a number of reports emerged claiming that the Iranian authorities promised to grant asylum and/or other socioeconomic benefits to the families of those Afghan refugees, who would join the Assad regime's forces as one of the primary conflict parties in the Syrian civil war.
2 Even though in many cases, official as well as unofficial ties to governmental structures exist, mostly in a form of personal ties or a regular business relationship.
3 More information at www.sonsoflibertyinternational.com/, www.americanmeso-potamian.org/ and www.1naef.com/.

References

Hegghammer, T. 2011, 'The Rise of Muslim Foreign Fighters. Islam and the Globalization of Jihad', *International Security*, 35(3): 53–94.
Hegghammer, T. 2013, 'Should I Stay or Should I Go? Explaining Variation in Western Jihadist's Choice between Domestic and Foreign Fighting', viewed 7 May 2017, http://hegghammer.com/_files/Hegghammer_-_Should_I_stay_or_should_I_go.pdf.
International Committee of the Red Cross. 2008, The Montreux Document, viewed 7 May 2017, www.icrc.org/eng/assets/files/other/icrc_002_0996.pdf.

Singer, P. 2007, *Corporate Warriors: The Rise of Privatized Military Industry*. Ithaca, NY: Cornell University Press.

Tonkin, H. 2011, *State Control over Private Military and Security Companies in Armed Conflict*. Cambridge: Cambridge University Press.

United Nations. 1977, Protocol Additional to the Geneva Conventions of 12 August 1949, and Relating to the Protection of Victims of International Armed Conflicts (Protocol I), 8 June 1977, viewed 7 May 2017, www.icrc.org/ihl/INTRO/470.

Williams, P. 2002, *Transnational Organized Crime and the State. The Emergence of Global Authority in Global Governance*. Cambridge: Cambridge University Press.

5   Uprisings, violence and the securitisation of inequality

Caroline Varin

Introduction

The 'Arab Spring' that began in Tunisia in 2011 alerted the world to the political impact of popular uprisings in the 21st century. A wave of protests that swept through North Africa has since moved south to Sudan, Burkina Faso and Burundi and spread east to Ethiopia and the Middle East. This has caused immovable governments to topple in some countries and grave uncertainty in others where little change has taken place since the start of the troubles.

The reasons that motivate thousands of citizens to mobilise against their political leaders are varied, but economic grievances are leading factors that explain the risks that people are willing to take when confronting the armed authorities (Cederman et al. 2013, Nagel 1974). The 'Arab Spring' has been explained as the natural consequence of a restless youth who, despite national growth and economic development, find a dearth of opportunities leaving them increasingly frustrated with a government that appears remote and stagnant. In a media-savvy world, anyone can now compare themselves not only to their neighbours but also to their leaders living extravagant lifestyles and people on the other side of the world who appear to have a better life than them. The gap between rising expectations and perceived reality inevitably increases the more we look at others, leaving a feeling of group discontent across all strata of society. The recent wave of demonstrations is also different from previous ones due to the active participation of 'urban and rural workers and the poor in general' (Gerges 2014), the citizens who had largely been overlooked by the political elite for decades. This may suggest that the uprisings were at least partly motivated by an overwhelming sense of discontent caused by existing and increasing inequalities rather than simply by economic factors such as poverty and youth unemployment.

In 2014 and again in 2017, the World Economic Forum (WEF) identified income disparity and attendant social unrest as the 'biggest single risk to the world' in its annual Global Risks Report (World Economic Forum 2014). This finding is repeated in the WEF 2015 Outlook Global Agenda (Global Agenda Councils 2015). Although the report does not find a direct correlation between inequality and political violence, it does link the vulnerability of disenfranchised populations to recruitment by terrorists, rebels and other armed groups that challenge the sovereignty of the state. The recent rise in violent non-state actors across the African continent and especially in countries that have experienced uprisings also suggests a correlation between failed economic aspirations and rebel movements.

Inequality has traditionally been recognised as a socio-economic or developmental problem, not as a potential threat to national security. Realising the impending problem, the governments of Nigeria and South Africa have, to a limited extent, focused on job creation and addressed some of the grievances of their population, although without really engaging in wealth redistribution campaigns. Other states have followed suit. However, the problem of inequality as a driving force in disintegrating societies is largely overlooked in these economic campaigns. This is particularly salient for a continent that hosts nearly all of the most unequal countries in the world, high levels of corruption and long-lasting illegitimate leaders, provoking the populace to seek redress through action rather than the ineffectual ballot box. This chapter argues that inequality has not been taken into serious consideration when assessing the drivers of the 'Arab Spring' and other uprisings. As a consequence, governments, especially in Africa, fail to address a possible root cause of popular discontent that will continue to haunt them for the foreseeable future. By focussing on inequality as a source of discontent and frustration, and as a motivator for mass political movements – but not necessarily violence as this is often triggered by the state itself – this chapter seeks to redirect research and policy towards addressing this pressing issue. It explores the possibility of securitising inequality as a potential mobiliser of economic measures and political will in order to urgently attend to this growing and potentially disastrous phenomenon.

Theory

Much of the literature on inequality is entrenched in the academic realms of political economy and development. Although inequality, or the distribution of 'extreme poverty and wealth', has been recognised as a catalyst for 'civil disintegration' and political violence since Aristotle and Plato (Nagel 1974, p. 453), there is little conclusive evidence regarding the relationship from a security perspective. As Cramer explains, 'sharply skewed income and wealth distribution does not always or even usually lead to rebellion' (Cramer 2005, p. iii) and as a consequence the 'linkages between economic inequality and violent political conflict' are unclear (Cramer 2003, p. iii). The threat of political violence is often anticipated by the government, which can respond using a variety of tools from repression/coercion (the coercive balance) to introducing 'pacification' policies such as food and fuel subsidies (Tilly 1990). As this chapter will show, the uprisings in Africa since 2005 started off mostly peacefully but turned violent in response to the use of force against protestors by the military and police. Fawaz Gerges (2014, p. 4) describes the movements as largely peaceful: 'on the whole, demonstrators behaved in a dignified manner, displaying a sense of solidarity with one another, a commitment to principled action and unity of purpose and ranks'. This does not support casualty reports for the uprisings however, unless the violence was largely perpetrated by the security forces (Human Rights Watch 2015). It is necessary to demonstrate, using discourse analysis in social media and news outlets, whether the violence in the African uprisings was in fact instigated by protestors or by the security forces (see Table 5.1) before considering whether any relationship exists between inequality and political violence. Indeed, if violence was triggered or escalated by the security state apparatus rather than the protestors, then it is impossible to claim a direct correlation between inequality and violent popular uprisings in Africa since 2005. This coincides in fact with much of the existing research on the matter (Nagel 1974; Cramer 2005).

Previous studies that have developed a methodology to measure a correlation between violence and inequality appear to have encountered similar problems, i.e., unreliable data due to poor records, especially in countries that experience both inequality and violence and therefore suffer from weak institutions and infrastructure. In addition, controlling for other independent variables is often a challenge, necessitating a thorough qualitative study in addition to the quantitative models. Zimmerman's comprehensive review of the literature for example reveals 'a linear positive relationship between socio-economic inequality and political violence' (Zimmerman 1980, p. 202), although he acknowledges the weaknesses in the methodology and data reliability in many of the existing studies. Nagel (1974), on the other hand, argues that the relationship is nonlinear. Muller (1997) explains that high levels of income inequality 'radicalise' and 'polarise' a portion of the population and finds that this contributes to instability in his cross-national study of 33 countries. A separate study of 71 developing countries by Alesina and Perotti (1996) found that income inequality coincided with social discontent and political instability but also to a rise in political assassinations. In order to establish the applicability of these results for the recent uprisings in Africa and the Middle East, it is important to first explore the instigator – whether this is the protest movement or the state security forces – of the violence as mentioned above.

According to Ted Gurr, the process of development, rather than inequality, is flagged as the better explanation for violent uprisings. Populations that experience a change in their economic and cultural environment – without the political adaptation necessary to reassure them and manage their frustrations – are more likely to lash out violently in what is referred to as the 'frustration-aggression nexus' (Gurr 2015). This frustration, according to our thesis, is born out of real and perceived inequality in African states. Other factors also play a part, along with inequality, in pushing people into the streets to demonstrate against the status quo. Hirshleifer (1995, 2001) points out the importance of opportunity, arguing that the severity of a population's grievances is not necessarily as influential as the perception that the time is right for rebellion. The perception that political uprisings had succeeded in overthrowing neighbouring dictatorships and had the attention and support of the West may have played a part in mobilising populations regardless of the level of discontent. Collier et al. (2004) in fact entirely disregard social concerns such as economic inequality as a motivator for rebellion, although this thesis does not seem to apply to political uprisings as they focus on a cost-benefit analysis of violent conflict. The same can be said of Cederman et al. (2013) who argue that grievances do in fact explain the outbreak of violent conflict – although as we show in this study, most of the violence is not triggered by the protestors but by the state.

The main difference between a civil war and a political uprising for our purposes is the militarisation of the former and its direct fatalities that far exceed the casualties of the latter (Kalyvas 2006). One could argue that a civil war is the militarised extension of a political uprising, as was evident in Libya and in Syria in 2011 when organised factions of the population took up arms in response to a government crackdown on protestors. This chapter is concerned primarily with exploring inequality as a root cause of mass mobilisation – which started off relatively peacefully in most African countries, rather than focus on civil war.

In addition, there is a cultural tolerance for a certain degree of inequality, according to Hirschman (1981), as people patiently wait for their turn to rise up. Grenier (1996) argues, however, that ideological movements are a substantial contributor to political uprisings as frustrations alone are generally not mass mobilisers. The role of new media during the Arab Spring in transmitting ideas of revolution across borders would therefore explain the domino effect of political uprisings across Africa and the Middle East in countries whose populations would not otherwise be 'ready' to publicly voice their economic grievances and frustrations.

Samuel Huntington warned in 1968 that 'where the conditions of land-ownership are inequitable and where the peasant lives in poverty and suffering, revolution is likely, if not inevitable, unless the government takes prompt measures to remedy these conditions' (Huntington 1968, p. 375). This chapter sets out to test whether there is any correlation between inequality and the uprisings in Africa since 2005. The focus on Africa specifically is natural considering that the so-called 'Arab Spring' has largely taken place in North Africa, starting with Tunisia and affecting many neighbouring countries over the past few years. In the event where a correlation can be found, then a case can be made for the securitisation of inequality. Securitisation here is understood as the process of transforming and addressing an issue not traditionally associated with security, such as poverty or inequality, into a matter of national security. In addition, the recognition that inequality can lead to political uprisings with dire consequences for the stability and security of the region, not to mention the survival of the state, may be the best motivator for leaders to prioritise a growing socio-economic problem.

The case for securitisation needs to be made with all due precautions however. Bozeman controversially argues that Western concepts of peace and security cannot apply to Africa, which has distinctive features due to the influence of indigenous culture on contemporary political perceptions of war and peace (Bozeman 1976). If Bozeman's thesis holds, it is doubtful whether policy makers in African states will be moved to act despite convincing evidence of a correlation between inequality and political uprisings. For the purpose of this study, we only focus on uprisings in Africa without seeking to draw comparisons with other regions. It therefore discounts much of the literature on uprisings that is imbedded in Western political thought, in particular, sources that focus on the importance of class and ethnicity as causes of political mobilisation.

Method

Existing research on political violence and inequality have faced the problem of data collection. There is no universal definition of inequality nor is there much reliable data for the majority of countries concerned with these problems. Indeed, it has been to the advantage of the elites in less egalitarian countries to bury the figures, if they even exist, and focus on other more obvious problems. Fearon and Laitin (2003, p. 20) lament the 'poor quality of inequality data', including the Gini coefficients published by the World Bank (2017) which is available only for a handful of countries and often does not show the change in inequality as the data is collected only once in a generation (if that). As Cramer (2005) points out, a lack of quality data does not mean that the hypothesis can be dismissed. Political uprisings may still be correlated with inequality, even if it cannot *now* be statistically proven beyond reasonable doubt. Indeed, some theoretical and empirical research mentioned above already suggests a convincing relationship between inequality and political violence, although this needs to be backed by data and tested with qualitative case studies.

This chapter seeks to answer the following two questions using quantitative and qualitative research:

- Is there a demonstrable relationship between the level of inequality and the prevalence of political uprisings in African states since 2005?
- In the case where political uprisings did occur, were these violent and who instigated the violence?

This will be assessed by comparing the data on inequality in countries that have experienced uprisings in the past ten years and those that have not (sample size $n = 42$). In particular, it would be interesting to assess whether there has been an increase over time in levels of inequality specifically in countries that have experienced political uprisings. Unfortunately, reliable data demonstrating a change in levels of inequality is difficult to obtain due to the lack of data collection in select countries. Alternatively, is it possible to infer a 'critical level' of inequality that may have contributed to the political uprisings by comparing the level of inequality for countries at the time of an uprising?

Depending on the data results, qualitative research will be required to test the nature of the political uprisings. Were they violent? If so, was this triggered by the protesters or by the state? If it is systematically the latter, then no relationship can be drawn between political violence and inequality in this study. In order to assess the origin of violence in the African uprisings, we have traced the earliest reports for each wave of demonstrations, referring to local and international news outlets that attribute the violence either to the state or to the protestors, and cross-referencing the reports using social media, in particular Facebook and Twitter.

This chapter does not seek to explain political uprisings solely through the population's frustrations with inequality. The purpose of this study is to demonstrate, if such a relationship exists, that countries with high levels of inequality are arguably more likely to undergo political movements that could threaten the stability of the country when the state responds with violence. The research therefore focuses on the more recent wave of political uprisings in Africa in the run-up to 2011 and thereafter. The countries included are drawn principally from Mampilly and Branch's (2015) list of states with significant protests since 2005 ($n = 39$), although the list is updated to include recent political uprisings in Burundi, and a number of test countries where no uprisings occurred ($n = 3$) out of a possible 52 African states. Mampilly and Branch include popular protests that involve significant society participation and are not focused narrowly on specific goals such as improved waves and human rights.

Table 5.1 shows that out of 98 incidents of protests, the government was the first to use violence in 60 per cent of cases. In 25 per cent of protests, the demonstrators themselves instigated the violence. In all other cases where

Table 5.1 Protests in Africa 2005–2015

	Protest	Gini Index (latest recorded)	Regime Change?	Violence? Gov or public?
Algeria	2011, 2012, 2013, 2014	29.0 in 2012	No	Yes, gov
Angola	2011, 2013, 2014	N/A	No	Yes, gov
Benin	2011, 2014	43.4 in 2011	No	Yes, gov
Botswana	2011	N/A	No	Yes, both
Burkina Faso	2014	N/A	Yes	Yes, public
Burundi	2011, 2014	N/A	No, attempted coup	Yes, public

	Protest	*Gini Index (latest recorded)*	*Regime Change?*	*Violence? Gov or public?*
Cameroon	2008, 2012	N/A	No	Yes, both
Central African Republic	2006, 2013, 2014	N/A	Yes (2013)	Yes, both
Chad	2010, 2014	43.3 in 2011	No	No
Cote d'Ivoire	2010, 2011	N/A	Yes (2011)	Yes, gov
Democratic Republic of Congo	2011, 2013, 2014	42.1 in 2012	No	Yes, public
Djibouti	2005, 2011, 2014	45.1 in 2012	No	Yes, both
Egypt	2008, 2009, 2010, 2011, 2012, 2013, 2014	N/A	Yes (2011-2012)	Yes, both
Ethiopia	2005, 2006, 2013, 2014	N/A	No	Yes, gov in 2005/06. No in 2013/14
Gabon	2009, 2011, 2012, 2014	N/A	No	Yes, both
Guinea	2007, 2009, 2013, 2014	33.7 in 2012	No	Yes, both
Kenya	2008, 2010, 2013	N/A	No (elections)	Yes, both
Lesotho	2011	N/A	No	Yes, gov
Libya	2011, 2013	N/A	Yes (2011)	Yes, gov
Madagascar	2009, 2010, 2013	N/A	No	Yes, gov
Malawi	2011	N/A	No	Yes, both
Mali	2012, 2013	N/A	Yes, Coup d'etat (2012)	Yes, rebels/gov
Mauritania	2011, 2012, 2013	N/A	No	Yes, gov

(*Continued*)

Table 5.1 Protests in Africa 2005–2015 (*Continued*)

	Protest	Gini Index (latest recorded)	Regime Change?	Violence? Gov or public?
Mauritius	2011	35.8 in 2012	No	No
Morocco	2011, 2012, 2013	N/A	No	Yes, gov
Mozambique	2010, 2012, 2013	N/A	No	Yes, both
Niger	2009, 2013, 2014	31.5 in 2011	No	Yes, both
Nigeria	2012	N/A	No (elections)	Yes, gov
Senegal	2011, 2012	40.3 in 2011	No	Yes, gov
Somalia	2010, 2014	N/A	No	Yes, rebels
South Africa	2009, 2010, 2011, 2014	63.4 in 2011	No	Yes, both
Sudan	2011, 2012, 2013	N/A	No	Yes, gov
Swaziland	2011, 2012, 2013	N/A	No	Yes, gov
Tanzania	2011, 2012, 2013	37.8 in 2011	No	Yes, both
Togo	2005, 2010, 2011, 2012, 2013, 2014	46.0 in 2011	No	No
Tunisia	2011, 2012, 2013, 2014	N/A	Yes (2011)	Yes, gov
Uganda	2011	42.4 in 2012	No	Yes, gov
Western Sahara	2011	N/A	No	Yes, public
Zimbabwe	2005, 2007, 2008	N/A	No	Yes, gov
Ghana		N/A	No	No
Namibia		N/A	No	No
Zambia		N/A	No	No

protests became violent, it was either unclear who had started the violence, or it was a combination of the government and the public. It also records too many vacuums in data collection with regards to inequality as defined by the Gini index. Indeed, most countries sampled did not have officially recorded inequality scores, and when they did, this was uniformly just one data point in the period between 2001 and 2015 making it impossible to track changes in levels of inequality. Where data was recorded, there was an overall average of 37.5 on the Gini index, where a score of 0 represents perfect equality and a score of 100 perfect inequality. A correlation between inequality and political protests cannot currently be established in this manner and therefore requires further quantitative and probably qualitative research.

On the other hand, it may be more beneficial to consider perceived inequality, which requires qualitative data and large N surveys to gauge the mood of the population involved in the protests. A report published by the World Bank in 2015 found 'significant differences between objective and perception data and between perceived and actual income distribution' (World Bank 2015, p. 22) in Egypt between 2000 and 2008. Indeed, one's perception of quality of life and expectations include factors such as public services, the quality of jobs available, security, the strength of institutions, access to justice and just remedy, corruption and political stability. Finally, expectations about the future are hard if not impossible to quantify. The World Bank's report on *Inequality, Uprisings and Conflict in the Arab World* concludes that 'the Arab Spring events appear to have been precipitated by broadly shared concerns that affected negatively the wellbeing of the middle class' (World Bank 2015, p. 28). Such a study is still missing for the African continent.

Costs of the African uprisings

Since 2005, uprisings in Africa have had real political and economic costs to the state, supporting the argument in favour of urgent measures via a process of securitisation. Four states have experienced political upheaval where the sitting president was ousted following substantial levels of violence. Tunisia, the first country to experience the uprisings, saw its long-term President Zine al-Abidine Ben Ali flee the country after weeks of unrest during which protestors clashed with the state security forces. In Egypt, President Mubarak stepped down following similar events that incurred over 864 deaths in a matter of weeks (Sadiki 2015). He was subsequently sentenced to life in prison while his successor, President Morsi, has been condemned to death. His counterpart in Libya, Muammar Gaddafi, met a violent end at the hands of rebel fighters. In Burkina Faso, another long-time leader, Blaise Compaoré was forced to resign and flee to Côte d'Ivoire after protestors stormed parliament in 2014. Despite attempts at democratisation and political normalisation, the aforementioned countries continue to suffer from high levels of political instability as the new leadership struggles with addressing the root causes of uprisings, allegedly inequality and other socio-economic grievances.

Other long-standing African presidents (and there are many), including Presidents dos Santos of Angola, Obiang of Equatorial Guinea, Biya in Cameroon, Mugabe in Zimbabwe, Museveni in Uganda, al-Bashir in Sudan, Déby in Chad, Afewerki in Eritrea and Jammeh in Gambia (until recently)[1], have been in – and abusing – power for over 20 years. That the success of the above-mentioned demonstrations in ousting unpopular rulers could inspire more uprisings against the state only strengthens the argument for leaders to address the root causes of discontent. Toby Dodge explains that 'the exclusion of the majority of the population from the economy' as 'family members of the ruling elite flaunted their wealth in the streets of Tunis and Cairo as standards of living for the majority of the population stagnated' exacerbated existing economic grievances and pushed the constituency into revolutionary mode (Dodge 2011, p. 10). The outlandish lifestyles of the Presidents and their cohort, in contrast to the lives of the populace, highlighted existing inequalities that needed just a trigger to push protestors into the streets.

In terms of financial costs, a report from HSBC bank published in 2013 predicted a decrease of 35 per cent in the output of countries that underwent severe political unrest during the 'Arab Spring'. This does not take into account the damage to infrastructure, lost foreign direct investment and reputational damage to businesses. Furthermore, 'weak economic growth and the disruption of tax collection mechanisms have weighed heavily on revenue, while increased debt-servicing costs and rising spending on subsidies and wages have driven expenditure upward' (HSBC cited by Smolinski 2013). In Libya, public revenue fell by as much as 84 per cent in two years (Smolinski 2013). Unemployment has risen by at least three percentage points in Egypt whereas measures of good governance and the rule of law have both diminished in the affected countries. All of this serves to undermine the government's spending power as its tax base shrinks leaving the state even more vulnerable to political unrest. In terms of human costs, 'the disruption of education, training and employment for scores of citizens will affect their ability to develop and contribute their human capital' (Jupp, cited in Sadiki 2015, p. 57).

In addition, the countries that have undergone a change of government in the aftermath of the protests have seen a rise in terrorist attacks and armed groups. Radical Islamist groups have been particularly successful in Libya and Egypt. Tunisia is considered to be the single largest supplier of ISIS fighters to Syria (Byrne 2014) and in 2015, the group released an issue of its magazine *Dabiq* dedicated to their expansion across Africa. The security implications for both the state and neighbouring countries, not to mention Europe just across the sea, are undeniable. This supports the World Economic Forum's urgent call to address inequality as a priority security threat. If inequality really is fuelling the terrorist recruitment drive, prompting uprisings and political instability, then the case in favour for securitising this problem is indeed conclusive.

The case for securitisation

Inequality has had a deeply disruptive effect on the social fabric of states around the world. Africa in particular has seen a rise in political uprisings in the last 10 years and since the 'Arab Spring'. In many countries, these uprisings have caused significant damage to the security of the state. In the last four years, Egypt's economy has been crushed by political uncertainty, mass mobilisation and an increase in terrorist attacks around the country. According to Khan writing about the Arab Spring, 'virtually all the economies have floundered over the past three years, experiencing both low economic growth and high unemployment' leaving the countries 'in worse shape than they were prior to the uprisings' (Khan 2014, p. 1). Khan points to the political turmoil and social unrest that created an environment of great uncertainty, which deterred investment, put off tourists and affected neighbouring states through a spill-over effect. The overall result has left countries in Africa even more fragile and vulnerable to political uprisings than before the turmoil even began.

A case for the securitisation of inequality can be built upon observance of the dire political and economic consequences that these uprisings have had in select African states. Securitisation, as defined by Buzan and Weaver (1998), is the subjective politicisation of a perceived existential threat thereby legitimising specific policies to address the issue. The politicisation of inequality mobilises the audience, in this case the state, which is persuaded of the severity of the situation and henceforth embarks on deploying emergency measures to counter it. Framing inequality as a security issue 'defines how we respond (...), how much is allocated in combating it and what sectors of the government are involved in the response' (Piot 2001, cited in Elbe 2009, p. 56). For something like inequality to be securitised, it needs to be understood as a pressing matter of national, regional and international concern, not just as a domestic socio-economic problem. Forcing a security label onto a non-traditional threat such as inequality is effectively a political choice, explains Weaver, but it is also a matter of responsibility. Securitisation 'serves to underline the responsibility of talking security, the responsibility of actors as well as analysts who choose to frame an issue as a security issue' effectively transferring it 'to the agenda of panic politics (rather than) handling it within normal politics' (Buzan et al. 1998, p. 34). Stephen Walt warns against expanding the field of security studies to include non-traditional threats, arguing that it risks destroying 'its intellectual coherence and make it more difficult to devise solutions to any important problems' (Walt 1991, p. 213). Buzan et al. indeed conclude that 'avoiding excessive and irrational securitisation is thus a legitimate social, political and economic objective of considerable importance' (Buzan et al. 1998, p. 208).

Just like HIV/AIDS, which was until recently portrayed as a developmental and public health issue before being securitised by the international community, inequality has security implications that transcend the state. Former World Bank president James Wolfensohn said in 2000 many of us used to think of HIV/AIDS as a health issue. We were wrong ... Nothing we have seen is a greater challenge to peace and stability of African societies than the AIDS epidemic ... We face a major development crisis, and more than that, a security crisis (in Elbe 2006, p. 121).

The disease was subsequently framed as a national security threat by the Clinton administration and has since been portrayed as a human security issue (Piot 2001, Leen 2004), a national security dilemma (Kristoffersson 2000, Sarin 2003, Ostergard 2005) and worthy of international attention due to its global implications (Prins 2004). Former CIA director George Tenet defended the securitisation of HIV/AIDS, explaining that the disease 'can undermine economic growth, exacerbate social tensions, diminish military preparedness, create huge social welfare costs, and further weaken already beleaguered states' (Tenet 2003). This echoes very closely with the impact that the political uprisings have had in Egypt, Tunisia and Libya, not to mention Burkina Faso and Ethiopia, among other Africa states that have experienced recent demonstrations against the state.

Securitising an issue however can have undesirable effects. Elbe (2006) shows in the securitisation of HIV/AIDS that the 'threat-defence' logic may mobilise the state into encroaching into the civil liberties of its population. In the case of inequality and uprisings, should a direct correlation be found, then the state could feasibly justify the use of force in an attempt to tackle socio-economic inequalities. Historically, the state has often taken advantage of so-called 'emergency' situations to silence any opposition and extend its own power over its citizens. It is therefore relevant to take into account the ethical considerations of securitisation when assessing whether or not an issue deserves to be moved into the agenda for security priorities. In many African countries however, the state is conspicuously absent from any kind of social services, which is a factor that exacerbates inequality, as the elite continue to accumulate opportunities while the rest of the population is left behind, unaided. Therefore, state mobilisation might only be possible if the issue is properly securitised, as security remains the primary concern of leaders. 'Appealing to the self-interest of states through the language of security can be economically useful' (Elbe 2006, p. 134) although a delicate balance must then be found to involve the state in addressing a pressing socio-economic issue relevant to its own security without tipping into an excessive militarised response driven by fear. Finally, securitising inequality could allow states to shift the responsibility for the relevant policies to political bodies with greater influence and a larger budget, thereby taking the necessary steps to address the problem of growing inequality among citizens.

Conclusion

Neglecting the real cost of inequality is dangerous. The disenfranchisement of a significant portion of the population, especially under-35s, from economic opportunities is arguably a more potent mobiliser of civil unrest that the exclusion of the population from political involvement. As Courtwright explains, 'whenever young, single men congregate for long period under other than stringent discipline, violence ensues' (cited in Tilly 2003, p. 1). The effect is multiplied when these men are unemployed and idle, leaving them open to indoctrination by armed groups with a political agenda, as evidenced in the successful recruitment strategy of ISIS/ISIL/IS in Syria and Iraq.

Nonetheless the securitisation of inequality warrants convincing evidence that the socio-economic problem is behind the mass movements rocking the continent. While this can be shown using data analysis and quantitative research, the questionable nature of much of the data collected in these countries undermines the reliability of our results. The World Economic Forum's Global Risks Report identified inequality as a security priority following interviews with 500–1,500 experts across the board. The qualification of inequality as a threat to international security by leading personalities around the world might not suffice to securitise the issue, but certainly raises an alarm that cannot be ignored.

Inequality as a socio-economic problem has appeared on the political agenda of the international community, with the creation of new institutions (International Inequalities Institute, LSE), the publication of books (Stiglitz 2012, Piketty 2014) and the release of reports (World Economic Forum 2014) that together show a growing awareness around the issue. This chapter has sought to contribute to the debate by focusing on the African continent. It argues that the political uprisings did not lead to violence, as the latter is most often perpetrated by the states themselves. Nonetheless, the political uprisings across Africa caused enough economic, political and social disruption to be classified as a serious security threat to the stability of the continent. We show that inequality, and especially perceived inequality, is an issue largely overlooked by academics and requiring further research, is a possible root cause of these uprisings. As a result, inequality needs to be approached through a security lens *as well as* through socio-economic policies.

Notes

1   It should be noted that protests in the Gambia in 2016 led to the resignation of President Jammeh in January 2017.

References

Alesina, A. and Perotti, A. 1996, 'Income distribution, political instability, and investment', *European Economic Review*, 40(6): 1203–1228.

Bozeman, A.B. 1976, *Conflict In Africa*. Princeton, NJ: Princeton University Press.

Buzan, B. and Waever, O. 1998, *Liberalism and Security: The Contradictions of the Liberal Leviathan,' Working Papers*, no. 23. Copenhagen: Copenhagen Peace Research Institute.

Buzan, B., Waever, O., and De Wilde J. 1998, *Security: A New Framework for Analysis*. Boulder, CO: Lynne Rienner.

Byrne, E. 2014, 'Tunisia becomes breeding ground for Islamic State fighters', *The Guardian*, 13 October, viewed 12 June 2015, www.theguardian.com/world/2014/oct/13/tunisia-breeding-ground-islamic-state-fighters.

Cederman, L.-E., Gleditsch, K. S., and Buhaug, H. 2013, *Inequality, Grievances, and Civil War*. Cambridge: Cambridge University Press.

Collier, P., Hoeffler A., and Soderbom, M. 2004, 'On the duration of civil war', *Journal of Peace Research*, 41(3): 253–273.

Cramer, C. 2003, 'Does inequality cause conflict?' *Journal of International Development*, 15(4): 397–412.

Cramer, C. 2005, 'Inequality and conflict a review of an age-old concern', *Identities, Conflict and Cohesion Programme Paper*, No. 11, Geneva: United Nations Research Institute For Social Development.

Dodge, T. 2011, From the 'Arab Awakening' to the Arab Spring; the Post-colonial State in the Middle East. LSE IDEAS Report.

Elbe, S. 2006, 'Should HIV/AIDS be securitized? The ethical dilemmas of linking HIV/AIDS and security', *International Studies Quaterly*, 50(1): 119–144.

Elbe, S. 2009, *Virus Alert*. New York, NY: Columbia University Press.

Fearon, J.D. and Laitin, D.D. 2003, 'Ethnicity, insurgency, and civil war', *American Political Science Review*, 97(1): 75–90.

Gerges, F. 2014, The Arab Spring Popular Uprisings—Myth and Reality, *Opendemocracy*, viewed 7 May 2017, www.opendemocracy.net/arab-awakening/fawaz-gerges/arab-spring-popular-uprisings-%E2%80%93-myth-and-reality.

Global Agenda Councils. 2015, *Outlook of The Global Agenda 2015*. Geneva: World Economic Forum.

Grenier, Y. 1996, 'From causes to causers: The etiology of Salvadoran internal war revisited', *Journal of Conflict Studies*, 16(2): 26–43.

Gurr, T.R. 2015, *Political Rebellion*. London and New York, NY: Routledge.

Hirschman, A. O. 1981, 'The changing tolerance for income inequality in the course of economic development', In A.O. Hirschman (ed), *Essays in Trespassing: Economics to Politics and Beyond*. Cambridge, England: Cambridge University Press.

Hirshleifer, J. 1995, 'Theorizing about conflict', In T. Sandler and K. Hartley (eds), *Handbook of Defense Economics*, vol 1. Amsterdam: Elsevier Science, 165–189.

Hirshleifer, J. 2001, *The Dark Side of the Force: Economic Foundations of Conflict Theory*, Cambridge: Cambridge University Press.

Human Rights Watch. 2015, *Egypt: Documented Death Toll From Protests Tops 300 | Human Rights Watch*, viewed 12 June 2015, www.hrw.org/news/2011/02/08/egypt-documented-death-toll-protests-tops-300.

Huntington, S.P. 1968, *Political Order in Changing Societies*. New Haven, CT: Yale University Press.

Kalyvas, S.N. 2006, *The Logic of Violence in Civil War*. Cambridge: Cambridge University Press.

Khan, M. 2014, *The Economic Consequences of the Arab Spring*. Washington, DC: Atlantic Council Rafik Hariri Center for the Middle East.

Kristoffersson, U. 2000, *HIV/AIDS as a Human Security Issue: A Gender Perspective*, paper presented at the expert group meeting on 'The HIV/AIDS Pandemic and Its Gender Implications', Windhoek, Namibia, 13–17 November.

Leen, M. 2004, *The European Union, HIV/AIDS and Human Security*. Dublin: Dochas.

Mampilly, Z.C. and Branch, A. 2015, *Africa Uprising*. London: Zed Books.

Muller, E.N. 1997, 'Economic Determinants of Democracy', In M.I. Midlarsky (ed), *Inequality, Democracy and Economic Development*. Cambridge: Cambridge University Press.

Nagel, J. 1974, 'Inequality and discontent: A nonlinear hypothesis', *World Politics*, 26(4): 453–472.

Ostergard, R.L. Jr. (ed). 2005, *HIV, AIDS and the Threat to National and International Security*. London: Palgrave.

Piketty, T. 2014, *Capital in the Twenty-First Century*. Cambridge, MA: Harvard University Press.

Piot, P. 2001, AIDS and Human Security, Speech Delivered at the United Nations University, Tokyo, October 2, http://data.unaids.org/media/speeches01/piot_tokyo_02oct01_en.doc

Prins, G. 2004, 'AIDS and global security,' *International Affairs*, 80(5): 931–952.

Sadiki, L. (ed). 2015, *Routledge Handbook of the Arab Spring*. New York, NY: Routledge.

Sarin, R. 2003, 'A new security threat: HIV/AIDS in the military', *World Watch*, (March/April): 17–22.

Smolinski, L. 2013, 'The Cost of a Revolution: $800bn', Financial Times, 10 October, viewed 12 June 2015, http://blogs.ft.com/beyond-brics/2013/10/10/the-cost-of-a-revolution-800bn/.

Stiglitz, J. 2012, *The price of inequality*. New York, NY: W.W. Norton & Co.

Tenet, G.J. 2003, *Testimony of Director of Central Intelligence George J. Tenet before the Senate Select Committee on Intelligence*. Washington, DC, February 11.

Tilly, C. 1990, *Coercion, Capital and European States, AD 990-1992*. Oxford and Malden, MA: Wiley-Blackwell.

Tilly, C. 2003, The Politics of Collective Violence. Cambridge: Cambridge University Press.

Walt, S. 1991, 'The renaissance of security studies', *International Studies Quarterly*, 35(2): 211–239.

World Bank. 2015, *Inequality, Uprisings and Conflict in the Arab World*. Washington, DC: World Bank Group.

World Bank. 2017, 'GINI Index (World Bank Estimate)', viewed 8 May 2017, http://data.worldbank.org/indicator/SI.POV.GINI.

World Economic Forum. 2014, *Global Risks 2014*, Geneva: World Economic Forum.

Zimmerman, E. 1980, 'Macro-comparative research on political protest', In T.R. Gurr (ed), *Handbook of Political Conflict: Theory and Research*. London: Macmillan.

6 Complicating security

The multiple narratives emerging from the Ukraine crisis

Crister S. Garrett

Opening observations

The unfolding security crisis in Ukraine has provided scholars with a rich (and unwanted) occasion with which to test and refine theories to explain international politics and security. Realists stress the role of material power, liberals the central role of institutions and constructivists the stable but dynamic politics framing decision-making. There are many veins of data and developments in the contemporary Ukraine crisis to mine and with which to either reaffirm or introduce generalised or more applied meanings. Scholars competing among relatively settled epistemological spaces bordering much of international studies scholarship have made fresh claims of legitimacy and seek to stake out further claims of ontological authority. Such exercises and intellectual competition are what make international relations as a field so stimulating, and challenging.

The debate has been so animated because the Ukraine crisis underscores how fluid and hybrid security politics has become since the Cold War. We are far removed from the seemingly settled post-Cold War order of the early 1990s. Indeed the Ukraine crisis confronts the transatlantic and international community fundamentally with fresh challenges for the protocols of war and the construction of peace and stability. The Ukraine crisis confronts Europe and North America squarely with the ethics of solidarity and the building of material, institutional and normative systems of security that are sustainable. Ukraine is, in short, a European, transatlantic and global laboratory for how the international community can confront not just the protocols of hybrid warfare, but also the protocols of hybrid diplomacy. The Ukraine crisis has complicated the understanding of security, and as such, opened spaces for the construction of new protocols of international order for the 21st century.

Transatlantic and global reactions to the Ukraine crisis as it has unfolded since February 2014 have clearly been coordinated internationally but simultaneously calibrated according to national or particular interests. One of the central and striking themes in the protocols of crisis management for the Ukraine crisis centers on German-American relations and the rapid developments in this relationship that have unfolded in the effort to coordinate transatlantic security politics. Europe's leading economic and arguably political power along with the West's leading economic, political, and military power have sought carefully to calibrate their public narratives offered to justify and legitimise Ukrainian policies made to date. As Chancellor Angela Merkel of Germany stressed during a press conference with President Obama in Washington D.C. on 9 February 2015, the United States and Germany are calibrating their Ukrainian policy 'with the closest of coordination' ('in engster Abstimmung'), and this is an important message for Russia to understand (President Obama and Chancellor Merkel 2015). For the German chancellor with an Eastern German biography, the West has arguably found a fresh and meaningful legitimacy in the Ukraine crisis with which to embed the ethical foundation of a transatlantic security architecture resting on the pillars of the European Union, NATO and the United Nations.

While the crisis has seemingly stabilised, if not further embedded a sense of security community among NATO and EU member states – a short-hand institutional and transatlantic definition of the West – that has not prevented multiple and indeed competing narratives within the transatlantic alliance to emerge with which to frame and pursue the politics of security in the context of the Ukraine crisis.

As Adler and Barnett (1998) have argued, competition in a security community is expected. And as Anderson et al. (2008) have noted, crisis can lead to a reaffirmed if not adapted community (or of course its dissolution). Competing and contested narratives certainly shape contemporary transatlantic politics. This was evident dramatically at the 2015 Munich Security Conference, a central arena for transatlantic security identity negotiation and construction (Bunde 2014). As Anne Applebaum (2015) from *The Washington Post* reported, the theatre of contrastive security philosophies and politics unfolded among transatlantic elites at the 2015 Munich Security Conference in rationally structured, emotionally driven performances that underscored basic elements of security policy convictions being challenged and defended. The formal protocol from the 2015 Munich Security Conference provides in essence a material, institutional and normative protocol of contrastive narratives of security, ethical assumptions about violence and order and ultimately the judicious path toward building peace.

Indeed, discernable at Munich and in the broader contours of the Ukrainian crisis are four identifiable narratives that have national anchorings but also transnational publics. These can be termed '*Ostpolitik* Redux', 'Progressive Realism', 'National Injury', and 'Enforced Order'. These narratives integrate realist, liberal and constructive elements to create critically-driven 'ecosystems of security'. Ecosystem is defined here as an observable space in which material, institutional and normative elements interact in an interdependent dynamic encouraging an ontological and epistemological personality or profile to emerge that informs (national and individual) political performance and policy frames. Security is defined as building on the traditional survival and stability of the state while integrating contemporary considerations of economic and identity wellbeing or a perception of relative situational stability. Critically-driven means an interpellative process whereby actors interact within a certain arena, ecosystem, or structure, by answering the call of a constructed ecosystem while practicing agency and thereby influencing the evolution of that ecosystem. The resulting ecosystems of security engender and embed what in turn can be referred to as 'cultures of security', i.e., practices, institutions and norms that are shaped and reinforced by a culture, in the anthropological meaning of the concept, that informs (and limits) the performance of security.

Focusing on German-American interactions in the transatlantic security community during the Ukraine crisis, this essay explores how connected but contrastive and indeed competing narratives of security are influencing and informing the Western ecosystem of security and thus shaping basic policy choices confronting respective governments and the international institutions in which they interact. The resulting fluidity in the Western ecosystem of security reveals new forms of 'complicating security' that portend possible evolutions in the relative state of the transatlantic equilibrium or an identifiable and functioning security community.

Sorting the narratives

Developments in Ukraine during the past two years have compelled national governments and key Western international institutions such as NATO, the EU, and the OSCE, to respond and to react. The German government has provided some of the clearest language and normative priorities for framing national and Western measures for containing but also resolving the Ukraine crisis. Chancellor Angela Merkel has provided unusually distinct borders for what she considers acceptable international behaviour. Speaking before the German parliament during the early phase of Russian involvement in the Ukranian crisis, she spelled out a diachronic definition of permissible security policy for Europe and for indeed a global system of order that must hold synchronic or contemporary meaning and purpose (Deutscher Bundestag 2014). Her convictions center on a modern international system of sovereign nation-states buttressed by international law that can build a thicker postmodern community of shared norms and institutions to enhance sustainable transnational security. Merkel's articulation of an international system of individual actors bounded and thereby liberated from insecurity with a structure of law-and-order illustrates the core belief in a global *Rechstaat* that defines a settled German narrative for a global *Ordnungspolitik*.

Scholars in the transatlantic community have themselves settled into the understanding that the traditions and practices of German security policy since World War II have created a national security culture framed by a post-Westphalian logic (Kirchner and Sperling 2010) that means formal and fuller concepts of sovereignty have been relativised by a constructive effort to build a defensive security architecture (Duffield 1999) with notable stability

(Berger 2003), making Germany largely a civilian power (Maull 1990–1991, 2000) with little to no interest in pursuing independent material power involving capacity to exercise violence in the international system. The thorough integration of German military structures into NATO and European security architecture underscores an institutional narrative of a bounded power.

Dialogue and diplomacy are the forms of security deployment that German elites and publics clearly prefer over traditional Westphalian logics of credible and sovereign material security. Realist predictions aside (Mearsheimer 1990; Waltz 1993), Germany has clung consciously to a carefully constructed postwar security culture deemphasising the possession and deployment of national power.

Traditions and narratives become all the more embedded with external, positive reenforcement. When German Chancellor Willy Brandt won the Nobel Peace Prize in 1971 for his administration's initiative known as *Ostpolitik*, it signalled for West Germans the arrival and acceptance of a German brand of security policy based on diplomacy and rapprochement, the deemphasis of material security and the recognition in the international community of a national contribution to a global *Ordnungspolitik* (Brandt is the only German since the Second World War to win the Nobel Peace Prize). A positive German brand had been established that helped distance the trauma of World War II and created externally legitimised societal space for a feeling of well-being (*Wohlgefühl*) around the pursuit of sustainable security in Europe. *Ostpolitik* was a regional narrative involving the Cold War and efforts at diplomacy and rapprochement, but for Germans and many others, it represented a global logic with direct implications for efforts at constructing global stability.

Practitioners of American security policy from John F. Kennedy to Richard Nixon to Ronald Reagan to Barack Obama could and can certainly understand the relative merits of diplomacy to pursue and maintain a liberal international order. The origins of the OSCE, with the establishment of the CSCE in the 1970s as part of the era of détente, reflected a process of encouraging Cold War stability that the Nixon administration actively pursued. The truism in American security culture that 'it took Nixon to get us to China' reflects an accepted logic, however that effective diplomacy and credible capacities in material security enjoy a symbiotic relationship. America's settled reading of détente stresses that the Soviet Union and China sought rapprochement because they had become convinced of America's 'staying power' and thus it would serve their interests to engage America's definition of a 'global *Ostpolitik*' to encourage a 'global *Ordnungspolitik*', assuming their international legitimacy and models for societal order. John F. Kennedy underscored in his inaugural speech that peace comes through strength; Ronald Reagan engaged Gorbachev because he was convinced American power had led the Soviet Union to seek rapprochement; and, Barack Obama's commitment to deploy substantially more American resources for NATO capacity in Poland and Eastern Europe reflects a symbiotic logic between defence and diplomacy that the president referred to earlier as part of an effort 'to raise the price for Putin's misconduct' (Remarks by President Obama and Chancellor Merkel 2015).

Of course, West German and German politicians and publics have long understood America's definition of effective diplomacy. Indeed the longing (*Sehnsucht*) for stability and order in the context of the Cold War and the post-Cold War era have led elites and voters in Germany to engage, if not embrace American material power embedded in NATO and deployed globally. Chancellor Konrad Adenauer's '*keine Experimente*' ('no experiments') message during the 1950s and 1960s captured West German sentiment very effectively, the desire for predictability and sustained security. Adenauer and the Christian Democrats enjoyed their degrees of postwar popularity and power based on this cautious or conservative logic. The practice of material credibility and its embeddedness in West German security culture was underscored when Social Democrat Helmut Schmidt, first as Defense Minister and then as Chancellor, called in the later 1970s and early 1980s for the West to counter Soviet missile deployments in Europe with their own 'peace through strength' policy, the so-called double-track strategy. Even after the Cold War, no German chancellor has questioned in any meaningful form Germany's membership in NATO or its centrality for a European and transatlantic security order.

The double-track policy of the early 1980s that Chancellor Schmidt played a central role in constructing was certainly embraced by the Reagan administration, but also rejected by many Americans. Indeed, the largest peace demonstrations in American history took place against this settled approach to security policy of 'preparing for peace by building strength'. Chancellor Kohl would be ultimately responsible for deploying the transatlantic double-track policy of threatening to deploy new medium range missiles unless the Soviets withdrew theirs from Eastern Europe. Massive demonstrations took place in Germany against the implementation of double-track, as they had when Adenauer announced West German rearmament and entry into NATO. Kohl would go on to serve as chancellor for another fifteen years, as did Adenauer for another eight years after West Germany became a full member of NATO. German and American security cultures have been contested and debated in short among national and transnational constituencies that have influenced substantially the transatlantic security community.

The collapse of the Soviet Union in 1991 seemed like a moment of international fluidity that could actually permit the dissolution of NATO to accompany the collapse of the Warsaw Pact. German and American voices argued that such a step could provide *Ostpolitik* new stature and influence, an opportunity to practice rapprochement with Russia and to reassure a country suffering from a deep sense of national loss and injury. Prominent American scholars and officials in the Clinton administration argued that, for Europe, peace through strength was no longer as necessary. Many voices in Germany supported this normative argument. A Kantian order of agreed upon norms had settled into Europe with which to construct a 'permanent' or sustainable peace. The dissolution of Yugoslavia and the resulting war in Europe, and Russian interests pursued in the conflict in the name of a pan-Slavic identity, complicated substantially such post-Cold War logics. During the 1990s, moreover, NATO continued to serve as a central institutional reinsurance for large European powers like France, Great Britain and Poland, that a unified Germany would remain a bounded power.

All participants in German, European and American debates about recalibrating post-Cold War narratives of security understood, however, that the theme of 'national injury' had to be taken seriously. The power of Nixon's détente policies toward China and the Soviet Union in the 1970s had been importantly about these 'Eastern powers' perceiving, via these initiatives, that they had 'arrived' in the global order, that they had become equals in standing with the West's reigning power. Insecurities and deep feelings of national injury from earlier Western incursions and invasions were thus relativised. Détente meant recognition by the West that China and the Soviet Union could enjoy a form of settled order among great powers based on mutually acknowledged stature. Détente resulted in a form of political pride in Beijing and Moscow. The collapse of the Cold War order shattered that pride in Russia, and provided new space for China to pursue more.

Related but distinctive narratives of security thus emerged during the Cold War that continue their trajectory involving Germany, Europe, the United States, Russia and shifting orders of global politics. What sociologists refer to as path dependency and anthropologists as cultures, can be perceived by international relations scholars as elements of realist, institutional and constructive logics integrating to encourage, embed and empower narratives of security that frame perceptions of international politics and resulting policy choice. The German-American security interdependence, interpellation, and, ultimately, intersubjectivity take place on a global stage with other key actors that challenge, recalibrate, and reenforce relative conclusions to new developments in international and global politics.

Gazing at the German-American debate about material, institutional and normative priorities for security policy, one can discern in a European and global context four identifiable narratives that have national anchorings but also clearly transnational publics and practitioners: '*Ostpolitik* Redux', or the reemphasis and refinement of this strategic diplomatic narrative; 'Progressive Realism', or the pursuit of peaceful change through strength; 'National Injury', or the striving for recognition through revisionist policies; and 'Enforced Order', or the search for stability by stressing *Rechtstaat* and the ability to uphold such *Ordnung*. Crisis or disruption complicates if not unsettles a relatively stable ecosystem in which publics and policy-makers frame and pursue choice.

The Ukraine crisis has provided arguably the most unsettling and disruptive development in the transatlantic security community within at least the last decade (post 9/11) if not quarter century (since the conclusion of the

Cold War). In this sense, the Ukraine crisis has fundamentally complicated the protocols of war and peace and confronted the international community with basic ethical choices about how best to encourage security, stability, and more profoundly, international solidarity.

The Ukraine crisis and negotiating narratives

Chancellor Angela Merkel's clear demarcation of acceptable international behaviour in her March 2014 Bundestag speech about Russia's invasion of Ukraine opened, simultaneously, space for the country to pursue an *Ostpolitik* by Germany's confident and centrist foreign minister, Frank-Walter Steinmeier. Clearly Germany's traumatic history with Russia during the 20th century compelled Germany to pursue the primacy of diplomacy (*Primat der Diplomatie*) to defuse the explosive situation in Ukraine. Both chancellor and foreign minister also understood that German voters longed for the strategy to succeed since neither they, nor the country's elites, wanted to consider the alternatives. Even possible costs associated with economic sanctions were enough to make a large majority of Germans against such measures initially. The original *Ostpolitik* of the 1970s had been, importantly, about opening new political spaces between the Soviet Union and the United States to pursue German security, defined in the 1970s rather as European and transatlantic interests as opposed to national interests. Thus was launched in the early months of 2014 Germany's first major 'shuttle diplomacy' initiative since World War II. Foreign minister Steinmeier travelled tirelessly between Kiev, Moscow, Berlin, Paris, and Brussels trying to coordinate, consolidate and ultimately construct a credible compromise for the Ukraine crisis.

Russian leader Vladmir Putin played along with these efforts for different reasons. Germany's economic investment in Russia is the largest of any Western country, and such material influence gives the country a unique place at the Kremlin power-table. Germany is also the most powerful economic and political power in the EU, and arguably the most powerful European member of NATO. Merkel had laid down, after all, the decisive veto against Ukraine's application to become a member of NATO being accepted at the Alliance's 2008 Bucharest Summit. If Germany were to veto Ukraine becoming a full member of the EU for similar reasons of not wanting to antagonise Russia, then Putin will have achieved at least one of his relatively clear goals, namely keeping Ukraine out of two key institutional pillars of the transatlantic or Western alliance. But for *Ostpolitik* to succeed, Merkel made clear in her Bundestag speech that it must involve the full territorial integrity of Ukraine, a guarantee that Germany strongly encouraged along with the signatory countries, the United States, Great Britain, and Russia, in the Budapest Memo on Security Assurances of 1994. And that promise no longer mattered to Putin 20 years later as he constructed a new narrative based on Novorossiya and built on Crimea, if not much more of Ukraine, being part of a Russian cultural space. Nurturing Russian patriotism and weakening a transatlantic security community have resulted in a rational Putin strategy of engaging German *Ostpolitik* while not committing to any of its long-term conclusions.

President Obama has supported Germany's contemporary *Ostpolitik* in the context of the Ukraine crisis for the same strategic logic that motivated John F. Kennedy to accept the Berlin Wall, for Richard Nixon to pursue détente and approve of Brandt's *Ostpolitik*, and for Ronald Reagan to reach out to Mikhail Gorbachev. In each case, a US administration, whether Democratic or Republican, saw the opportunity as less costly to defuse international tension, and encourage regional and international stability, thus distancing America from having to become more involved in European security mainte-nance if not enforcement. Detailed accounts of Obama's approach to the cur-rent Ukraine crisis underscore that the American administration has eagerly accepted German leadership (Financial Times 2015). As with Kennedy, Nixon and Reagan, Obama has sought German answers to European security crises because America's overall global security obligations make the option hard to resist for avoiding, or at least limiting, increased commitments to new regional challenges where a politically reliable and materially rich ally is available.

What is arguably qualitatively different during the Obama administration is the president's longer term security strategy of openly allowing Germany and Europe to emerge as clearly more independent regional and global actors. Brandt travelled to Washington DC to get Nixon's approval for *Ostpolitik* (for which the original thinking emerged after Brandt, as Lord Mayor of West Berlin, had been so bitterly disappointed in Kennedy's quick acquiescence to the Berlin Wall), Reagan's team made clear who was in charge in the Western alliance, and even the Clinton administration struggled with how far Europe should create an independent security profile within the transatlantic com-munity. Few such qualms or calculations have impressed themselves upon the Obama administration. As with his predecessors, Obama's strategic logic has been imposed upon him due to the relative limits, if not relative decline, in American material resources. More philosophically, Obama has been guided by a strategic vision calling for the progressive empowerment of regional allies that can contribute to an emerging system for global order. From the 2009 London G20 Summit where Obama came 'to listen and not to dic-tate' to Libya and 'leadership from behind' (Obama Press Conference 2009), Obama was crafting a progressive realism allowing for more alliance agency to empower a sustainable security architecture for Europe.

Obama's strategic pragmatism and confidence has been matched by Merkel's moral clarity, resulting in *Ostpolitik* not only having new space to be practised but for its calibration to new realities. As it has become increas-ingly clear that Putin was essentially playing along, or playing for time, the German government has recalibrated the relative benefits of the primacy of dialogue and diplomacy, increasingly willing to practise material power with economic sanctions and fresh commitments to NATO and actual German military capacity. Foreign minister Steinmeier in his own confidence and strong adherence to the transatlantic alliance has also not been unwilling to relativise the centrality of *Ostpolitik* for his own agenda, and that of his party, the Social Democrats. Germany's new regional and global stature and Steinmeier's ambition add up to a willingness in the German government to reach sober conclusions about Putin's agenda in Ukraine. Merkel is not disinclined to encourage that process, having reportedly pushed for the Russian leader's permission to have Steinmeier sit in on meetings the two held at the November 2014 Brisbane G20 Summit so he could experience first-hand how Putin consistently avoided articulating any sort of commitment to resolving the crisis. (Financial Times 2015). Such a learning process has meant that German *Ostpolitik* has been repositioned importantly from an *Ostpolitik* as a positive good in and of itself to an *Ostpolitik* Redux whereby the primacy of diplomacy has been enhanced with the integration of a clearer role for peace by strength, or more pointedly, via economic sanctions and a buttressing of NATO hard capabilities. As Chancellor Merkel underscored during her press conference with president Obama on 9 February 2015 in Washington D.C., her administration agreed fully with the American logic that the material price for unacceptable international behaviour must be made clear and indeed raised if necessary, in short, punishing unacceptable forms of behaviour.

The act of publicly punishing bad behaviour can of course trigger different responses. And the motivations for such behaviour in the first place can have multiple causes. Russia's 2014 invasion of Crimea and Ukraine enjoyed deep popularity at home, especially among the Russian middle class. Putin was showing the world that his country could 'stand up' to the West, 'defy' the West, 'impose new conditions' on the West. It was arguably the most empowering moment in contemporary Russian history after the trauma of the Soviet Union disappearing from modern maps. Such a vocabulary of standing up, defying, and imposing has driven much of the post-Cold War recalibration of global power politics in the context of a new global *Ostpolitik*. While a key feeling involved is of resentment and injury, it is interrelated naturally with a national longing for pride and a secure feeling of patriotism or national identity. This accounts importantly for why the Chinese middle-class stands so strongly behind the country's ambitious agenda in the Pacific region and globally.

The original détente and *Ostpolitik* of the 1970s had been rather about arrival and recognition; contemporary *Ostpolitik* as practiced by Beijing and Moscow is more about revision and recalibrating global order. These may be elite-constructed narratives, but they enjoy strong popular support, tapping into longer-term traditional narratives of national injury.

Russia and China are certainly not alone in engaging such feelings to construct national security narratives. The United States has of course been haunted and driven by the aftermath of 9/11 and the need to reassert national agency and confidence in purpose. Within Germany itself, many political and popular voices have empathised with Putin's narrative of being deceived by the West, of being lied to by the West, of the West imposing its power, as they struggle with their legacy of being an occupied country for so long. Especially in Eastern Germany, with its complicated narratives emerging from national unification and the feeling among many that a West German model of societal order was imposed with little voice of their own (Bittner 2014), there is widespread sympathy for Putin's politics of injured pride and how it plays into and reconfirms an East-West binary. Of course Putin's recent maneuverings in Eastern Europe have triggered fresh rounds of national fear and reserve among purported allies as well. Thus, formal members of his security initiative, the Eurasian Economic Union, have to date not formally recognised the Russian take-over of Crimea, and neighbour Belarus has resurrected its formal economic border with Russia (including switching back to carrying out economic transactions in euros and even dollars, and not just rubles). There is plenty of national injury to go around that is informing European and transatlantic politics around the Ukraine crisis. The challenge with feelings and international relations of course is that they introduce fluid dynamics that make security ecosystems notably less stable and complicate the space between definitions and perceptions of rational and irrational behaviour. For America and its political traditions of poetic politics (New Morning in America, Yes We Can) and ambitious, visionary politics (send a man to the moon), there is considerable room for such imaginary politics. But for Germany and its traumatic 20th century experiences with poetic politics and visions for world order, there is much less inclination (but certainly fascination) to pursue inspired politics. The German political class would much rather inform interests with institutional and finely calibrated definitions of incremental and hopefully sustainable political and security order. Such cautious or conservative instincts lead Germany's centrist politics to remain clearly distanced from any politics of resentment (or ressentiments) and to instead seek to establish stability by inter alia deliberating between injured parties or playing the widely-accepted role in Germany of arbitrator, or *Vermittler*.

Indeed German media, politicians, and publics have embraced the image of Germany as *Vermittler* in the Ukraine crisis between the 'emotions' of Russia and America, two competing great powers in many Germany eyes that do not share a German definition of *Rechtstaat* and rational behaviour. What is for Germany a rational response triggers however strong emotions and memories elsewhere. In Poland, for example, part of national trauma is Germany and Russia soberly negotiating the fate of Central Europe 'over their heads'. This has not been the case at all in the current Ukraine crisis, indeed, Poland has gained new stature in Europe with its forceful efforts to find a solution to the crisis. Yet Poland has also been a strong advocate of peace-by-strength and a progressive realism for European security. Chancellor Merkel has thus been careful not to raise her profile as a *Vermittler* between Russia and Europe where Poland might not have a formal seat at the table (for example, at the so-called Minsk II negotiations). At a moment of interesting international political theatre, the Merkel-Obama press conference on 9 February 2015, video of the press conference shows the chancellor about to use the word *vermitteln*, to arbitrate or mediate, but stops herself and rather stresses Germany's role in 'standing up for European order'. It is a snapshot of the power between language, logics and the construction of international political narrative (C-SPAN 2015).

The complications for Germany, the United States, Europe and the transatlantic community are indeed to be found in Merkel's words 'standing up for order' and what this entails for national and international policy. Prominent voices in Germany have even argued that it should involve the country supporting the eventual supply of defensive weapon systems to Ukraine. The President of the Munich Security Conference, Wolfgang Ischinger, one of Germany's most distinguished (retired) diplomats having served as ambassador in the United States (at the time of 9/11) and Great Britain, has argued that in a situation with 'no good solutions', Germany must be also willing to assist Ukraine independence with military support (Ischinger 2015). Even foreign minister Steinmeier has voiced the sentiment that if the so-called Minsk II accord (February 2015) does not hold, Germany should be ready to debate this option. And while Chancellor Merkel stated in her press conference with President Obama in February 2015 that she sees 'no military solution', what was rather surprising was the absence of her categorically excluding the option of defensive weapons systems being delivered. She did stress that she agreed fully with America's strategic logic that, if needed, the price for Russian aggression in Ukraine must be raised.

On the flip side, prominent American policy makers and scholars have argued forcefully against supplying weapons to Ukraine. Eugene Rumer, a senior adviser in the Clinton, Bush, and Obama administrations, has said such a move would risk another 'Black Hawk Down' scenario, a reference to the American catastrophe in Somalia (Rumer 2015). And policy veterans at the Brookings Institution, a 'training ground' for Democratic administrations, have engaged in an animated and public political debate about the relative merits of supplying weapons, with Brookings president Strobe Talbott, a seasoned 'Russia hand', calling for such a move, while other voices have argued decidedly against such a logic (Talbott and Pifer 2015; Shapiro 2015). Prominent realist scholar John Mearsheimer has also argued forcefully against sending weapons into the region, maintaining that the West has already contributed substantially to worsening the conflict, if not actually causing it (Mearsheimer 2015, 2014).

President Obama himself was very careful during his 9 February 2015 press conference with Chancellor Merkel to stress that no decision has been made whether to actually ship defensive weapons to Ukraine, that it remained a real but unrealised option. That still remains the case over a year later, but the 'technology' or material option remains just as potent too, and the ethics behind it still intact. Further complicating the transatlantic debate about 'enforcing order' are the diverse European and North American voices that differ strongly on the material upholding of a transatlantic security community beyond the current economic sanctions and enhancement of

NATO capabilities. Canada has agreed with the United States, for example, that all options should be considered, including defensive weapon systems being supplied to the Ukrainians. The Italian, French, and Hungarian governments have spoken out clearly against such an option. Great Britain and Poland have spoken forcefully to consider and indeed prepare such an option. The thickness or durability of the transatlantic security community and its spaces for democratic deliberation can be witnessed in the open and contested debate about what enforcing order or upholding a European and transatlantic security community should fully entail. The Ukraine crisis and the policy issue of defensive weapons systems underscores the potency and implications of the relationship between technology, ethics, and the protocols of war and peace.

Conclusions: Complicating narratives

The Ukraine crisis has illustrated that there are at least four discernible narratives of security binding and complicating the transatlantic security space. These are '*Ostpolitik* Redux', 'Progressive Realism', 'National Injury' and 'Enforced Order'. Both German and American debates about how to frame the Ukraine crisis and form policy choices contain elements from each narrative while also allowing for discernible and distinctive national narratives. Germany's initial and still prominent narrative focuses on Ostpolitik, but is undergoing important recalibrations. America's 'Progressive Realism' still shapes the primary national narrative but is being contextualised by what President Obama refers to as 'strategic patience', or a cautious approach to enforcing order. Both Berlin and Washington D.C. are engaged – especially in the personalities of the chancellor and president – in a robust learning community where both parties are carefully calculating their relative strengths and how to coordinate their deployment to resolve or at least to stabilise the Ukraine crisis and its regional and global implications.

Beyond bilateral relations, both Germany and America have embraced transatlantic and European institutional tools with which to construct both national and transnational policy for the Ukraine crisis. At the September 2014 NATO Summit in Wales, resolutions were reached to prepare a rapid deployment force for Eastern Europe. The resulting Dutch-German Corps forms the basis for this material strengthening of European security enforcement that will also include a supply-and-training center in Poland (Szczecin). The July 2016 NATO Summit in Poland witnessed Germany and the United States agreeing to substantial new alliance forces being stationed in Poland and the Baltic states, a proactive policy that would have been unimaginable prior to the Russian invasion of Ukraine. At the same time, no other country has called more strongly for the OSCE to play a monitoring and assessment role in Ukraine than Germany. And the EU has played the central role in coordinating European sanctions strategy against Russia, and in turn, being the lead voice in coordinating sanctions strategy with the United States.

Thus, multiple and related material, institutional, and normative elements are being remixed and repositioned within the transatlantic security community to encourage the evolution of distinct but interdependent security narratives. The resulting evolution in turn of the transatlantic security space or ecosystem will not only influence regional and transnational security politics and policy for the forseeable future, but will certainly shape global debates, politics, and power as the international community continues to struggle with constructing a function system of global order. That effort is being centrally informed, evident in the Ukraine crisis, by the inherent relationship between technology, ethics, and the protocols of war and peace.

References

Adler, E. and Barnett, M. 1998, *Security Communities*. Cambridge: Cambridge University Press.

Anderson, J. J., Ikenberry, G. J., Risse, T. (eds). 2008, *The End of the West? Crisis and Change in the Atlantic Order*. Ithaca, NY: Cornell University Press.

Applebaum, A. 2015, 'The Long View with Russia', *The Washington Post*, 8 February.

Berger, T. U. 2003, *Cultures of Antimilitarism: National Security in Germany and Japan*. Baltimore, MD: Johns Hopkins University Press.

Bittner, J. 2014, 'Eastern Germans' Soft Spot for Russia', *International New York Times*, 31 December.

Bunde, T. 2014, *Transatlantic Identity in a Nutshell: Debating Security Policy at the Munich Security Conference, 2002–2014, Transworld Working Paper no. 45*.

C-SPAN. 2015, President Obama and Chancellor Merkel Press Conference, 9 February, viewed 30 June 2017, https://www.c-span.org/video/?324265-1/president-obama-german-chancellor-angela-merkel-news-conference.

Deutscher Bundestag, D. 2014, 'Tagesaktuelles Protokoll der 20. Sitzung vom 13. März 2014: Abgabe einer Regierungserklärung durch die Bundeskanzlerin zum Treffen der Staats- und Regierungschefs der Europäischen Union zur Lage in der Ukraine am 6. März 2014'.

Duffield, J. S. 1999, *World Power Forsaken: Political Culture, International Institutions, and German Security Policy after Unification*. Palo Alto, CA: Stanford University Press.

Financial Times. 2015, 'Battle for Ukraine: How the West lost Putin', *Financial Times*, 2 February.

Ischinger, W. 2015, 'Pledge Weapons for Ukraine of the Violence Will Go On', *Financial Times*, 5 February.

Kirchner, E. J. and Sperling, J. (eds). 2010, *National Security Cultures: Patterns of Global Governance*. London: Routledge.

Maull, H. W. 1990–1991, 'Germany and Japan: The New Civilian Powers', *Foreign Affairs*, 69(5): 91–106.

Maull, H. W. 2000, 'Germany and the Use of Force: Still a "Civilian Power"?', *Survival*, 42(2): 56–80.

Mearsheimer, J. J. 1990, 'Back to the Future: Instability in Europe After the Cold War', *International Security*, 15(1): 5–56.

Mearsheimer, J. 2014, 'Why the Ukraine Crisis is the West's Fault', *Foreign Affairs*, 93(5): 77–89.

Mearsheimer, J. J. 2015, 'Don't Arm Ukraine', *International New York Times*, 9 February.

Obama Press Conference. 2009, 'Obama Press Conference, London G-20 Summit', 2 April, viewed 14 July 2016, www.cbsnews.com/news/transcript-obamas-g20-press-conference/.

President Obama and Chancellor Merkel, 2015, 'Remarks by President Obama and Chancellor Merkel in Joint Press Conference 2015, The White House, Washington DC.', 9 February, viewed 30 June 2017, https://obamawhite-house.archives.gov/the-press-office/2015/02/09/remarks-president-obama-and-chancellor-merkeljoint-press-conference.

Rumer, E. 2015, 'Arm Ukraine and You Risk Another Black Hawk Down', *Financial Times*, 4 February.

Shapiro, J. 2015, *Why Arming the Ukrainians is a Bad Idea*. Washington, DC: Brookings Institution, 3 February.

Talbott, S. and Pifer S. 2015, *Arming the Ukrainians and Russia's Reactions.* Washington, DC: Brookings Institution, 5 February.

Waltz, K. N. 1993, 'The Emerging Structure of International Politics', *International Security*, 18(2): 44–79.

7 Technology, development, global commons and international security

A global commons and interdisciplinary approach to global security

M. Matheswaran

Introduction

The concepts of national and international security have necessarily become far more complex in the rapidly changing and globalising world of the 21st century. The world today is ever more interdependent than one would have imagined even half a century ago. However, security as a concept still continues to be heavily influenced by traditional threat and response mechanisms. Traditionally, security is seen as an outcome of effective management of the three pillars of power: economic, military, and political instruments of national power.

> Economic power is derived from the resources and technology within a state's borders and the aptitude to trade them; military power comes from the availability of people, material, and technology; and political power is drawn from the potency of leaders and institutions, the people's support and endorsements from other nation-states (Dunn Cavelty 2007, p. 89).

In the past, conflicts have been the results, invariably, of intense competition amongst the states for resources and acquisition of territories. Wealth creation and, hence, increase in the power of a state, has always been seen as resulting from its ability to conquer territories and thus gain control of resources. The results of these, in turn, were meant to create advantages in trade and domination of markets. Industrial revolution and technological growth gave rise to colonial powers, which exercised enormous control over vast geographical and manpower resources. The end of the colonial system after the World Wars and the end of the Cold War in 1991 have resulted in a world of more than 200 independent, sovereign nation-states and, for most of them, values of democracy, self-determination and sovereignty have become the bedrock of international relations. Ideology-based territorialism, and hunger for resource-driven territorial expansion, in the colonial mould, is largely irrelevant in the 21st century. New patterns are emerging for territorial control, such as the religious fundamentalist route by ISIS for control across multi-states, and the historicism route established through reclamation by China in the South China Sea. However, nations continue to be driven by the objectives of wealth creation. Various methodologies are followed in pursuit of this objective: free flow of trade, technology domination, creation and domination of markets through product-driven strategies, and domination of the global commons for distinct advantages to those nations who have started early and have dominant technologies. It gives rise to classical resource-based competition amongst states, which is a serious source of conflict with adverse impact on international security.

The world has changed considerably over the last few decades, primarily due to the impact of the digital technology. As the Tofflers' have highlighted, the world is in a transformation or 'power shift' (Toffler 1990), from industrial age economy to knowledge age economy. At the core of this transformational process lies the challenge of dealing with authority and security, which continues to be dominated by the realist school. The invention and subsequent rapid growth of the Internet, global communications and greater integration of the world after the Cold War has accelerated the process of globalisation and interdependency. In turn, competition manifests as a threat to universal access to global commons such as the seas, space and technology. This creates huge problems for development and peace. In such an environment, solutions to security challenges lie, probably, in adopting a globally common, integrated approach, or, in short, a global commons approach.

Globalisation and security in an interdisciplinary world

The world today is far more integrated and interdependent than ever before. It began to shrink as technological revolutions in communication and transportation began to sweep the world from the late 19th century. With the rapidity of technological growth in ICT (Information and Communications Technology) and space-based services, the world has become even smaller. Technology, education, democratisation, global flow of trade, commerce, and human resources and the interdependent nature of the world economy have all necessitated an interdisciplinary perspective of the world in dealing with international affairs. The resulting globalisation imposes a multidisciplinary perspective when dealing with security in the 21st century.

Globalisation: concept and reality

After the end of the second World War, the world rapidly slid into an ideological contest and, through it, into two rival camps. Power and national security were viewed as inevitable factors in a zero-sum game. Hans Morgenthau's concept of ruthless realism, or classical realism, tempered by the practical structural realism of Waltz as the sole guiding principle of national security policy, was a given. Robert G. Patman states that this state-centred approach dominated international relations from 1945 to the end of the Cold War and was characterised by the core belief that international security is essentially defined by the military interactions of sovereign states. Patman (2006, p. 16) observes that this realism failed to predict the end of the Cold War and the subsequent break-up of the Soviet Union in 1991. He further observes that the end of the Cold War resulted in structural changes in the international system, and this, in turn, has raised a big question mark over the traditionalist, realist concept of security. But this development has been evolutionary.

In the immediate aftermath of the Cold War there has been a greater emphasis on power politics and a realist approach to national security in the context of the then-emerging American unipolar dominance. Beginning with the US-led intervention in Iraq in 1990–91, the technological and military superiority of the USA became the centrepiece of global security politics with the objective of shaping the world to enhancing unipolar dominance and convenience. However, this strategy began unravelling in the emerging complexities that impacted international security: the rise of religious fundamentalism, technology denials and its counterproductive effect on development, information technology and its huge impact on global connectivity, criticality of trade on global interdependency, emerging powers and the rise of China, human rights, competition for dominance of the oceans and space, environment and climate control and the emerging conflict between development needs and environmental protection. An era of technology and trade-enabled globalisation began to unfold from the early 1990s. However, the 9/11 attack on the World Trade Center in Manhattan renewed emphasis on a distinctly state-centric approach to security, including the war on terror. Large international backing for the American unilateral militaristic approach to the war on terror began to fall apart as the USA extended its applicability to areas of its own narrow interests rather than a global interest. American intervention in Iraq in 2003 was done without UN Security Council approval as the US-led coalition of the willing, a dozen US allies, showed scant regard for international opinion and norms while dismembering Iraq. Subsequent developments in Iraq and Afghanistan proved the ineffectiveness of a militaristic approach to complex security problems of the 21st century world. Following events highlighted the power of ethnic groups and the difficult power matrix that determines stability in diversity. In an unfolding era of 'deepening globalisation' the American unilateral militaristic approach has been a singular failure. The social unrest and 'Arab Spring' effects that have put vast regions of Syria, Iraq, Libya, Egypt, and Yemen, in major turmoil for many years, exemplify the complex nature of international security. Modern technology has aided the terrorists and revolutionaries, as much as it has the governments, in connectivity and enhanced awareness.

In order to understand the complexity of 21st century security dynamics and its relevance to global commons, one needs to grasp the concept of modern globalisation in correct perspective. Patman uses an interesting approach to define globalisation as 'Essentially Contested Concept' (Gallie 1962, p. 121). Is that a contest of time and space? Cultures and civilisations have interacted since time immemorial to have globalisation in various forms enabled by trade and economy. However, modern day digital and satellite technologies have enabled global connectivity and enhanced awareness. Effectively, it has been an intensification of interconnections between societies, institutions, cultures and individuals on a worldwide basis. In short, globalisation implies a shift in geography whereby borders have become increasingly porous (Scholte 2001, p. 14). Ian Clark (1997, p. 15) finds globalisation as a process that involves compression of time and space. Access to people and places through modern digital communications has resulted in dramatic reduction in the time taken, either physically or virtually, to cross borders and boundaries. As a consequence, the world is perceived as a smaller place as issues of environment, economics, politics and security intersect more deeply at more points than previously was the case (Patman 2006, p. 4).

The demand for an international and integrated approach to resolve global issues stems from the view held by a significant majority that the traditional state-centric silo mentality is anachronistic in an environment of interdependent economies. For some time now, the view has gained ground that growing interconnectedness of national economies through globalisation gradually negates the significance of territorial boundaries (Held and McGrew 1999, p. 4). It led to the premature ideas of the demise of the sovereign nation-state but by the end of the first decade of the 21st century, scholars have come around to hold a view that the rigid concept of the all-powerful nation-state is giving way to a flexible and adaptable nation-state that retains some aspects of its power, but needs to integrate significantly into taking a community approach to addressing various issues. In a world where security is and will always be the prime concern of states, strong proponents of globalisation advocate an interdisciplinary approach to security. They expect globalisation to gradually reduce and ultimately eliminate the space for states to manage national security policy. Where borders are becoming increasingly fluid and particularly non-existent in an increasingly powerful cyber world, national governments of the future will have little choice but to accept that the security agenda is shifting from one centred on the military capability of the state towards that of common or cooperative security. Examples of this approach, symptomatic of a move away from the sanctity of the state as the central focus in security concerns, are steps taken by the international community for humanitarian interventions in the recent past, like in Somalia, Bosnia and Kosovo. However, biased interventions in cases like Iraq in 2003, Libya, interference in Syria and failure to prevent genocides in Rwanda and Sri Lanka significantly weaken the credibility of the international community. Sceptics continue to propagate the state-centric approach. This approach is championed in most cases, by countries like Russia and China. Given the present dynamics, there are mounting transnational pressures on the sovereign state from within and without, necessitating a broader and more cooperative approach to security.

More importantly, globalisation has brought to the fore the importance of a cooperative approach to resolving problems related to global commons. This has pitted the development needs of developing countries against the conflicting needs of global environment protection. These involve various domains: technology, space, oceans and maritime freedom, environment and green house issues, renewable resources, etc. Each of these has implications for security and hence, globalisation necessitates a cooperative and interdisciplinary approach to security.

Security and the state

If globalisation has given rise to an appreciation of the multi-dimensional nature of security, then the evolving concept of security in an era of globalisation demands a careful analysis. The concept of security continues to be rooted in a state-centric approach in realist traditions with the foundational premises of the anarchic nature of the international system and the inevitability of war. The structural realism of Kenneth Waltz (*Man, State, and War*), where anarchy in the system influences behaviour of states, as against Hedley Bull's society of states, as much in power struggle and balance of power machinations, are influenced by interdependence, institutions and norms (*Anarchical Society*) will continue to characterise the realist spectrum of the 21st century state, even though increasingly challenged by globalisation. This traditional view of security becomes relatively simplistic when compared to the complexities that have arisen in an era of globalism where exponential growth of technologies, information and communication technologies in particular, introduces many dimensions to security.

Even though a traditional state-centric approach to security was manifestly focused on military power, it had its linkages to other instruments of national power – economic, technological and political. To that extent, in a simplistic manner, it was multi-dimensional or interdisciplinary. For a state to build significant military power, it needed wealth; economic power or wealth was attained through increasing trade and domination of world markets. But economic power needed the backing strength of military power to create a safe and secure environment in which trade and economic activity could flourish. Military power is becoming more and more technology intensive and generally functions at the top end of technology. Effectively, technology determines a state's success in capturing markets, and hence, it is now a very critical attribute of economic and military power. Only smart and visionary political leadership can harness this complex interdependence between technology, economy, military power and trade. Bertrand Russell articulated, decades ago, the relevance of economic war potential.

Traditional state-centric security strategies have continued, if not became stronger, in the bipolar Cold War period from 1945 to 1991. Many events that emerged simultaneously since the late 1980s – end of the Cold war, technology explosion aided by ICT, rapid globalisation aided market forces and an integrating world economy – have all put the state's role in security into a new perspective. Steve Smith (2006, pp. 33–55) analyses the role of the state in national security through traditional and alternative models:

1 The traditional view of security is a realist model, where the state's place in the international system will dictate its security and foreign policies. Globalisation has necessitated modifications to the traditional model whereby the state adapts to the demands of a cooperative international approach but nevertheless retains its focus on a militaristic security strategy. Calls for widening the scope of security to address issues like environment have increased since. Smith cites four groups who attempt to modify the traditional view – the first group sought to redefine the concept of security away from focus on military interactions; the second group views and analyses the concept of security from the interests of the rest of the world other than the great powers, fundamentally it looks at weaker nations' problems emanating from the challenges and barriers to their development, where security relates largely to internal challenges as opposed to the focus on external challenges for developed nations; the third group deals with a holistic, sociological approach where state identity is diluted in the context of global interests; and the fourth group represents 'neoclassical realism' where security and foreign policies are governed by domestic factors such as political and economic ideology, socio-economic structure and national character.

2 Smith examines six alternative approaches to security, of which only three are significant. The first is the constructivist security studies in which the state actors might see security as achievable through community rather than through power. Security, therefore, is something that can be constructed – a constructivist approach that recognises the 'importance of knowledge' for transforming international structures and policies. The second is a combination of the security approach of 'the Copenhagen School' and the 'Human Security' approach endorsed by the United Nations. Here the concept of security is viewed more holistically. Barry Buzan, one of the proponents of the Copenhagen School, broadened the security agenda to involve five sectors rather than the traditional focus on only one of these, military security. Buzan (1983, 1991) added political, economic, societal and ecological security sectors, while still keeping the state as the referent object of his analysis. Combining with Ole Wæver and Jaap de Wilde in their 'securitisation' concept, Buzan brings in a community and an interdisciplinary approach, calling it 'societal security' (Buzan et al. 1991, 1998; Wæver et al. 1993; Wæver et al. 1989). The broader concept of human security emerged out of the 1994 United Nations Development Programme (UNDP), and was followed up in 2000 with 'millennium development goals' with a target date of 2015. Essentially this is no different from the Copenhagen School approach and outlines seven areas of human security: economic security,

food security, health security, environmental security, personal security, community security and political security. The essential difference is that the UNDP focuses on the individual while Barry Buzan and others continue to keep the state as the central actor. The UN identified six main threats to human security: unchecked population growth, disparities in economic opportunities, migration pressures, environmental degradation, drug trafficking, and international terrorism (United Nations Development Program 1994). The third alternative approach that merits consideration flows from 'critical security studies'. It has a postmodern world approach to security, where the main focus is on emancipation. Its proponents, as Smith identifies, stress the need to move from a focus on the military dimension of state behaviour under anarchy to a focus on individuals, community and identity (Krause and Williams 1997).

Smith clearly establishes the fact that there are strong voices like Emma Rothschild to widen and deepen the concept of security. However, he also identifies the dissenting voices that strongly oppose any dilution of the concept of security from its focus on the military dimension. According to Smith, 9/11 has only encouraged hardliners and proponents for American supremacy to strongly support an American hard-line strategy of focusing on the military dimension and enhance its areas of influence.

Intersection of globalisation and security

The end of the second World War ushered in the demise of the colonial empires, and gave rise to a number of new nation-states. However, ideologies continued to hold sway on world politics, thus resulting in a bipolar world, locked in a Cold War, in the post-1945 era. The end of the Cold War not only ended the bipolar world but also ended the neatly divided stable world on ideological lines that had subsumed the cultural, social and religious identities and sovereign characteristics of a large number of states. However, the world continues to function on a realistic model where the anarchic nature of the world is its defining characteristic. The structure of the world continues to be arranged as per the interests of a handful of powerful states. In the contemporary context this is a dynamic and a very complex situation. Nowadays, globalisation and security intersect to make a powerful need for coordination. Technology, particularly ICT, has made borders meaningless in the cyber world and has allowed for the flow of ideas and knowledge across nations and cultures.

Globalisation is largely driven by technological progress and economic integration. Using the principle of networking, globalisation can be explained as a short hand for an array of phenomenon that derive from unorganised and stateless forces but that generate pressures that are felt by states. It is defined well by the Deputy Prime Minister of Singapore, Lee Hsien Loong, thus

Globalisation, fostered by free flow of information and rapid progress in technology, is a driving force that no country can turn back. It does impose market discipline on the participants, which can be harsh, but is the mechanism that drives progress and prosperity (Kelly and Olds 2005, p. 1).

Global communications have enabled rapid economic growth in developing countries. The last two decades have been characterised by a significant increase in the volume of international trade, especially in the scale and mobility of investment capital. This expansion in trade and mobility of capital is further underlain by phenomenal increases in speed of communications and transportation that have literally shrunk the world in time-space compression, which is one of the most important characteristics of globalisation. The impact of time-space compression exceeds the economic realm. It results in profound changes in the way people in different parts of the world view themselves, their futures, and the ways in which they are, in turn, impacted by developments in distant places as a result of such compression.

It becomes clear that technology and economics create an intersection of globalisation and security, which in turn, breaks the barriers imposed by traditional security thinking. Jagdish Bhagwati puts it in the following perspective –

> economic globalisation constitutes integration of national economies into the international economy through trade, direct foreign investment (by corporates and multinationals), short-term capital flows, international flows of workers and humanity in general, and flows of technology.

As a result, in a globalised world, economic management, decision making, production, distribution and marketing are organised on a global scale, which limits the nation-state's ability to regulate its own economic interests, and makes national welfare heavily dependent on the international market (Ripsman and Paul 2010, pp. 6–7). Globalisation, enabled by technology, has broken national barriers and forces the world to look at issues in an integrated and globalised outlook. Nations are forced to look at security from a global perspective, where their policies can be questioned from a perspective of their ability to help the cause of global public goods.

As a result, the intersection of globalisation and security, as Martin Shaw (2000, p. 6) points out, renders territorial boundaries irrelevant. In his opinion, it nullifies the cultural, political and technical boundaries that defined distinct worlds, isolated some social relations from world markets and inhibited communications. Kenichi Ohmae contends that under the irreversible influence of modern information technology, genuine borderless economies are emerging (Ohmae 1995, p. 7; Friedman 2005). Susan Strange (1996, p. 4) finds that the state, as the sole authority on security, is in retreat. Where states were once the masters of markets, now the markets, on many crucial issues, are the masters over the governments of states.[1] Gilpin (2000, p. 18) concludes that, in a highly integrated global economy, the nation-state has become anachronistic and is in retreat. Myriam Dunn Cavelty highlights the critical importance of the information infrastructure of the cyber world in integrating the world and hence, its role in the security architecture. Using chaos and complexity theories, Dunn Cavelty (2007, pp. 85–105) brings out the increasing complexities involved in addressing different dimensions of security in the digital world. She uses Moore's and Metcalfe's laws to highlight the increasing complexity in information infrastructure systems and growing interdependencies. Nations have come to depend greatly on ICTs in various sectors such as information and telecommunications, financial services, energy and utilities, transport and distribution. Consequently, these complexities and extensive ICT influence on critical infrastructures have given rise to enhanced security threats and vulnerabilities in the cyber world. An attack on any infrastructure has a 'force multiplier' effect, thus allowing even a relatively small attack to achieve much greater impact.

Thus, the intersection of globalisation and security highlights the fact that security threats emanate from a variety of technology-related factors, with ramifications across boundaries. Solutions to these will be outside the capability of state-centric security policies. Domains such as technology, space and cyberspace can only be handled through an integrated and cooperative approach. International security, then, needs to be viewed from a multidimensional approach involving much of the global commons.

The concept of commons, known and discussed for more than a century, involves those assets that are seen as common heritage of mankind. The term 'global commons' evolved as a collective label for the areas of 'Antarctica, the high seas and deep sea minerals, the atmosphere and outer space'. The rationale for combining these four physically distinct entities under the rubric of global commons stems from their shared attribute of being 'resource domains to which all nations have legal accesses' (Buck 1998, p. 1). Hence, the logic is that these should be preserved and kept free of competitive approach by nations to ensure that all of mankind is free to use them without hindrances. Free of competitive approach means keeping global commons, such as space, free of power politics and thus free from any potential military conflict. The reality, however, is quite different. International and multilateral institutions manage global commons. Ideally, these institutions would be representative and unbiased. Since the international system continues to be anarchic, the great powers would tailor the institutions and the use of global commons to maintain their status quo. Indirectly, the same power-politics afflict various policies with respect to the use and administration of global commons. For example, exploitation of space shows how great power rivalries continue to afflict this global commons and has already militarised it.

Global commons include space, oceans, environment, technology, cyberspace, climate and renewable resources such as water, marine life, etc.

This study looks at three of the domains: technology, space and cyberspace and examines the need for a common approach to security.

Commons – A basic approach

The problem of commons was recognised early when Katherine Coman highlighted its economic importance more than a century ago, in 1911 (Stavins 2011, pp. 81–108). As the industrial revolution based economies grew, their rapacious consumption-based developments have had an impact on the natural resources and environmental quality. Problems associated with management of open access and common property resources have been severe. With the developing world adopting a leapfrog strategy, the problems have become worse. These relate to air and water quality, hazardous waste, species extinction, maintenance of stratospheric ozone and most recently, the stability of the global climate in the face of the steady accumulation of greenhouse gases. Unregulated and open access to renewable resources has resulted in major problems to security – an example is the degradation of ocean fisheries. Similarly, the degradation of environmental quality is the direct result of unchecked and competitive consumption, manufacturing and trade. Development of a market-based approach to environmental protection, including emission taxes and trading rights, is a workable solution. These have potential to address the ultimate commons problem of the 21st century, global climate change (Stavins 2011, p. 82).

Security view of the commons

Great powers of the international system have always worked to keep the commons free, primarily from a military point of view to enable easy access to their military forces. The larger objective, of course, is to ensure their freedom to trade and access markets. As Mahan observed, the British Empire was built and endured because of its mastery of the seas. In a similar way, the USA's strategy of keeping the global commons accessible to all is fundamentally focused towards ensuring freedom of action to its military. The adverse fall out of this is that competitive power politics enters the commons domain. The US sees the global commons as the essential conduits of US national power in a rapidly globalising and increasingly interconnected world. Its Quadrennial Defence Review Reports assign 'assured access' to the commons a top priority for the US military forces (Redden and Hughes 2010). A natural reaction for this approach is similar individual state-centric strategies pursued by other powers, China, Russia, India, etc. For example, China's declaration of 'ADIZ' of the airspace over Spratly Islands in the South China Sea is a clear violation of the international norms on global commons. China's action adversely impacts the sovereign rights of the countries affected – Vietnam, Philippines, Indonesia, and Malaysia, etc. A security approach in the classical great power format would not be effective in addressing the challenges that would afflict the global commons in the 21st century.

Technology

Technological development has been instrumental in defining the modern world and will continue to define the postmodern world as well. While the developed nations achieved their current prosperity on the strength of their advantage of driving the industrial revolution, they have now leapfrogged into the knowledge world. Technology has been intimately linked to trade markets and so, economic development and has also been at the core of competitive politics. Two-thirds of the world missed the industrial revolution on account of being colonies of the European imperial and colonial powers. During this period two-thirds of the world were resource providers (raw materials) and, hence, were peripheral to the world economy while the colonial powers held complete control over technology and manufacturing. While colonialism ended long ago, most of the world continues to use technologies they do not own.

Since the beginning of the nuclear age, technology controls have been well-refined to ensure they remain restricted to the developed countries. Technology denial regimes such as MTCR, NSG, etc. restrict free flow of most technologies of today. As a result, advanced research is not possible in most countries due to denial of access to databases. The classification and restriction of dual-use technologies hinders or slows down the economic development of most countries. Technology denial is an extremely important and powerful tool in the hands of few countries. The biggest impediment to international security and development remains these instruments of power politics.

One of the critical benefits of the information revolution is the access to knowledge for anyone in the world through interaction across boundaries. While advanced technologies are safeguarded, the world is going through a knowledge revolution as a result of the process of technology diffusion. Sharing of knowledge on science and technology through combined research across national boundaries is a fast-growing process. National security mechanisms will have only limited capability to prevent this flow.

As science and technology spreads across the world new challenges to international security will arise. These emanate largely from transnational and non-state actors and terrorists who will also have the benefit of modern technology to aid their strategies. In the light of these developments global politics of science and technology can be focused towards making societies better. The challenges, as mentioned earlier, can only be tackled on a global approach and universal standards of humanity. It needs a two-pronged approach.

The first is to use technologies for economic progress and improvement of human conditions. It is assumed that a combination of science and technology offers better solutions to critical global challenges such as security, public health, energy, food and water supply, poverty and climate stability. The second is to focus on access to more advanced technologies that support the trend towards more efficiency, smartness and artificial intelligence to enable society improvements in areas of urban planning, reproduction, business models, etc. towards fostering wealth and a decent life for nine billion people.

International institutions such as World Bank and different UN initiatives call for closer international cooperation in scientific research and technological development. There is a need for global development research that makes use of open data, open access to research pools and collaborative knowledge production. India has been at the forefront of these efforts when it introduced the draft resolution 'The Role of Science and Technology in the context of International Security and Disarmament' in 1988. The UN General Assembly passed the resolution 43/77A on 7 December 1988, directing the UN Secretary General to follow future S & T developments, especially those which had military applications, evaluate their impact on international security, and to submit a report to the General Assembly in 1990. This has remained on the agenda ever since.

Space

Nowadays it is inconceivable to think of everyday life without the use of space assets and services. Space-based services such as satellite television, mobile services, Internet, global navigation, maps, weather forecast, telecommunications, disaster and rescue management, urban planning and many others have become part and parcel of everyday life. Space-based C4ISR is critical to all military operations. In a networked environment, denial of space services would create a huge imbalance between the two forces in conflict. Space input is critical to generating real-time situational awareness for military commanders. The quality of this service will directly impact the command and operations. Space, therefore, has become extremely vital to all nations, major powers in particular.

The launch of Sputnik in 1957 by the USSR initiated the space race. Both powers initiated their developments with military objectives. Science and exploration followed later. The 1967 Outer Space Treaty signed by all major powers acknowledged the sanctity of space as a global common and hence, preserve it as a common heritage for mankind. Notwithstanding this agreement, the leading powers have continued to militarise space, while adhering to the stipulation of not placing a WMD in space. Most space technologies are dual-use technologies. Space provides an arena for intense military competition. Space technologies are critical for military operations in terms of intelligence, precision weapons, communications, navigation and weather forecasts. As long as these remain militarily significant, the danger of space becoming an arena for military operations is high.

While leaders USA and Russia have become more cooperative than competitive, China continues to take on a competitive course with an aggressive approach. Although both USA and Russia competed fiercely during the Cold War period, subsequent cooperation on the International Space Station is an excellent example of the global commons approach (Sheehan 2007, pp. 174–81).

Space is critical to international security. If well-organised and managed by a transparent international mechanism, space could prove to be a model of enabling security using the global commons approach. However, there continue to be huge challenges associated with it. At the political level lies the mistrust amongst nations. China has clearly stated that it does not subscribe to the global commons approach unless all its interests are met. As a rising superpower, China could play the spoilsport. By testing of ASAT capability in 2003 the PRC clearly demonstrated its classic realist approach. More than 1200 satellites orbit Earth, providing tangible economic, social, scientific and strategic benefits to billions of people. The biggest threat to these benefits is the increasing number of debris that is created by individual countries (Secure World Foundation 2014). Space Situational Awareness (SSA) is an extremely important component of space capability. It is technology intensive and is used selectively. Unless the military capabilities in space are made transparent, steps towards implementing a global commons approach would become difficult.

Cyber world

Next to space, the cyber world portends unlimited capabilities. The cyber world is truly a globalised world with enormous potential for destructive warfare, irrespective of one's physical strength, size or wealth. Hence, the importance of cyber security comes into critical focus in the 21st century. While the world is becoming increasingly dependent on ICT for governance, banking, infrastructure, security, intelligence, welfare, etc., cyber terrorism is a capability that looms large on various nations. Cyber terrorism is defined as 'unlawful attacks and threats of attack against computers, networks, and the information stored therein when done to intimidate or coerce a government or people in furtherance of political or social objective' (Denning 2000). While a cyber terrorist threat is obvious, states use much of the same methods against victim states. The Stuxnet attack that immobilised the Iranian nuclear reactor presumably involved both Israel and the USA. The Russian attack on Georgia in 2008 was preceded by massive cyber attacks that neutralised that nation's information infrastructure. China employs upwards of 100,000 hackers in cyber attacks all over the world. A major objective of this strategy is acquisition of scientific research, technical and defence data on a large scale. For a country that has technology leapfrogging as its strategy, this is an important conduit through which it acquires its database.

Currently, the international system has not addressed the issue effectively. The taxonomy of cyber threats involves threats to the Internet infrastructure, threats to individual networks or servers, and threats to critical infrastructures. Of these, attacks on infrastructure could be lethal. Cyber threats have the potential of spinning out of control. How are cyber attacks by nations to be construed? Would it fall within the ambit of violation of the state's sovereignty? If so, does that amount to an act of war? A global commons approach will need clear definitions on these issues, lay down rules of engagement, create adequate transparency and work in a cooperative mode to address the threats. Cyber terrorism must be dealt with by multilateral and cooperative organisations.

Conclusion

The concepts of globalisation and security are as old as mankind. However, these two concepts assume different dimensions in the 21st century. For the first time ever in human history, mankind and the Earth have been integrated and brought together as never before in time and space. Information and communication technologies have created a world in which knowledge and connectivity are the driving factors. When space and aviation technologies are integrated with the ICT, the capabilities increase enormously. If states use these capabilities in a classical, realist theme, then national and international security would be much harder to achieve. The silver lining is that technology and globalisation have ensured that security can no more be only military focused, but must encompass many dimensions and hence, be interdisciplinary.

Economic and technological interests drive globalisation and it is irreversible. Similarly, security concepts are forced to become more holistic and include many dimensions, including development and human security. In the 21st century, globalisation implies de-territorialisation or the end of geography, which means the core basis of state power has eroded. States, therefore, have to address their security needs in an integrated and cooperative manner through the involvement of regional and global institutions.

Nevertheless, it is a reality that development is not uniform. In an incisive analysis, Robert Cooper finds that the current world exists in three levels – pre-modern, modern and postmodern. He finds the European Union as the postmodern state where national territories are integrated, states function independently while simultaneously being open to intrusive verification, particularly of their military capability, and the economic union is complete and foreign policies are increasingly integrated.

The EU will be hard to achieve for a long time in rest of the world. In the meantime, the only way to achieve reliable international security lies in adopting a global commons approach to the most important commons of mankind – space, technology and cyberspace.

Note

1   For example, one of the crucial restrictive factors that limited the expansion of the 1999 Kargil War between India and Pakistan was the important influence that business houses like Reliance wielded in the government to prevent economic damages to global business interests. Reliance's oil refinery on the Gujarat coast, one of the biggest in the world, is too close to the border and would have been the first to be hit in the event of a full-scale war.

References

Buck, S. J. 1998, *The Global Commons: An Introduction*. Washington, DC: Island Press.

Buzan, B. 1983, *People, States and Fear*. Brighton: Harvester Wheatsheaf.

Buzan, B. 1991, *People, States and Fear: An Agenda for International Security Studies in the Post Cold War Era*. 2nd ed. Hemel Hempstead: Harvester Wheatsheaf.

Buzan, B., Kelstrup, M., Lemaitre, P., Tromer, E., Wæver, O. 1991, *European Security Order Recast: Scenarios for the Post-Cold War Era*. London: Pinter Publishers.

Buzan, B., Wæver, O., de Wilde, J. 1998, *Security: A New Framework for Analysis*. Boulder, CO: Lynne Rienner.

Clark, I. 1997, *Globalisation and Fragmentation*. Oxford: Oxford University Press.

Denning, D. E. 2000, *Cyber Terrorism: Testimony before the Special Oversight Panel on Terrorism*, Committee on Armed Services, U.S. House of Representatives, May 23, viewed 15 May 2015, www.terrorism.com/documents/denning-testimony.shtml.

Dunn Cavelty, M. 2007, 'Securing the Digital Age: The Challenges for Complexity for Critical Infrastructure Protection and IR Theory', In J. Eriksson and G. Friedman, T. L. 2005, *The World is Flat: A Brief History of the Twenty-First Century*. New York, NY: Farrar, Straus and Giroux.

Friedman, T. L. 2005, The World is Flat: A Brief History of the Twenty-First Century. New York: Farrar, Straus and Giroux.

Gallie W. B. 1962, 'Essentially Contested Concepts', In M. Black (ed), *The Importance of Language*. Englewood Cliffs, NJ: Prentice Hall, 121–146.

Gilpin, R. 2000, *The Challenges of Global Capitalism: The World Economy in the 21st Century*. Princeton, NJ: Princeton University Press.

Held, D. and McGrew, A. 1999, *Global Transformations*. Cambridge: Polity Press.

Kelly, P. F. and Olds, K. 2005, 'Questions in a Crisis: The Contested Meanings of Globalisation in the Asia-Pacific', In K. Olds, P. Dickens, P. F. Kelly, L. Kong, H. W. Yeung (eds), *Globalisation and the Asia-Pacific*. London and New York, NY: Routledge, 1–15.

Krause, K. and Williams, M. 1997, 'From Strategy to Security: Foundations of Critical Security Studies', in K. Krause and M. Williams (eds), *Critical Security Studies*. Minneapolis, MN: University of Minnesota Press, 33–60.

Ohmae, K. 1995, *The End of the Nation State*. New York, NY: Free Press.

Patman, R. G. 2006, *Globalisation and Conflict: National Security in a New Strategic Era*. London and New York, NY: Routledge.

Redden, M. E. and Hughes, M. P. Nov 2010, 'Global Commons and Domain Interrelationships: Time for New Conceptual Framework', *Strategic Forum*, SF No. 259, Washington, DC: National Defence University, INSS.

Ripsman, N. M. and Paul, T. V. 2010, *Globalisation and the National Security State*. New York, NY: Oxford University Press.

Rothkopf, D. 1997, 'In Praise of Cultural Imperialism', *Foreign Policy*, 107, 38–53.

Scholte, J. 2001, 'The Globalisation of World Politics', In: J. Baylis and S. Smith (eds), *The Globalisation of World Politics: An Introduction to International Relations*. Oxford: Oxford University Press, 13–34.

Secure World Foundation. 2014, *Space Sustainability*, viewed 15 May 2015, www.swfound.org.

Shaw, M. 2000, *Theory of the Global State: Globality as an Unfinished Revolution*. Cambridge: Cambridge University Press.

Sheehan, M. 2007, *The International Politics of Space*. London and New York, NY: Routledge.

Smith, S. 2006, 'The Concept of Security in a Globalising World', In R. G. Patman (ed), *Globalisation and Conflict: National Security in a 'New' Strategic Era*. London and New York, NY: Routledge, 33–55.

Stavins, R. N. 2011, 'The Problem of the Commons: Still Unsettled after 100 Years', *The American Economic Review*, 101(1), 81–108.

Strange, S. 1996, *The Retreat of the State: The Diffusion of Power in the World Economy*. Cambridge: Cambridge University Press.

Toffler, A. 1990, *Powershift: Knowledge, Wealth, and Violence at the Edge of the 21st Century*. New York, NY: Bantam Books.

United Nations Development Program. 1994, *Human Development Report 1994*. New York, NY: Oxford University Press.

Wæver, O., Buzan, B., Kelstrup, M., Lemaitre, P. 1993, *Identity, Migration and the New Security Agenda in Europe*. London: Pinter.

Wæver, O., Lemaitre, P., Tromer, E. 1989, *European Polyphony: Perspectives beyond East-West Confrontation*. Basingstoke: Macmillan.

8   Organisational networks in post-conflict disarmament efforts

Andrea Kathryn Talentino, Frederic S. Pearson and Isil Akbulut

Recent trends in international responses to domestic civil war and humanitarian crises highlight the growing importance of multilateral intervention by intergovernmental organisations (IGOs). Organisations at regional, extra-regional and global levels have been involved in coincident or collaborative operations, both with other organisations and with individual nation-state and non-state interveners, which opens a set of compelling questions about the interrelations and networking behaviour of IGOs. Academics and practitioners agree that higher levels of collaboration among organisations should yield better outcomes, and entities such as the US Army's Peacekeeping and Stability Operations Institute (PKSOI) have devoted much effort to facilitating and improving such collaboration (Cimbala and Forster 2010). Yet the specific topic of interorganisational relations in humanitarian interventions remains largely uncharted, despite its growing importance for the future of international peacemaking and peace building.

Although there is a growing focus on organisational network interactions in many disciplines, much of it has taken place outside the field of political science. The networking literature within the field emphasises non-governmental organisations (NGOs) and private negotiators, rather than IGOs (see Atouba and Shumate 2010; Garb and Nan 2006; Nunez and Wilson-Grau 2003; Liebler and Ferri 2004), which provides an important but incomplete picture of organisational networking habits and impacts. Because IGO memberships are composed of nation-states, they are shaped by specific interests, agendas and policy concerns that are less relevant in the NGO context. This makes their interactions a compelling area of study, but thus far there have been few systematic efforts to examine how IGOs coordinate activities or the factors that enable or inhibit effective collaboration, mutual learning and positive outcomes.

The issue of effective collaboration is particularly important in the difficult and delicate tasks of implementing peace agreements and undertaking protracted multilateral interventions in the aftermath of civil war. Disarmament and demobilisation efforts, which the UN describes as 'crucial components of both the initial stabilization of warn-torn (sic!) societies as well as their long-term development' (United Nations 2016), are not only essential for moving affected societies into a post-conflict phase, but, as seen in instances such as Kosovo and Liberia, can be a major political obstacle. Disarmament of the Kosovo Liberation Army (KLA), for example, was part of the NATO-brokered and enforced agreements to end the fighting in that province and uphold Kosovar security *vis a vis* its neighbours. Yet the KLA evidently retained arms for some two years after the agreement, partially under reformation as a proto-national guard force. Splinter groups may have retained arms for even longer periods.

Post-conflict disarmament is a complicated and cross-cutting issue demanding consideration of multiple social factors including ethnicity, gender, age and community, as well as multiple political factors such as rule of law, security, corruption, territorial claims and economic development. Calibration of effort across all these factors is required for success in conflict resolution, placing emphasis on the interactions among and collaborations across IGOs, as well as NGOs and states that often contribute to multilateral efforts. Thirty years ago Christer Jonsson (1986) proposed the use of interorganisational theory for understanding the general topic of IGO interaction, though so far, few studies have followed in that vein. In this study we will build on Jonsson's analysis in order to analyse the emergent ramifications of organisational networks and their impact on the implementation and effectiveness of disarmament, demobilisation and reintegration (DDR) efforts in post-conflict interventions.

We analyse the disarmament and security strategies proposed or implemented by different organisations, as well as the extent of collaboration with other agencies involved in the same or complementary activities, to assess the factors that impede or enable effective collaboration, mission achievement and sustained peace agreements. In recognising the complexity of this analysis, we present this chapter as a first attempt in laying the groundwork for examining interorganisational networks in both process and outcome.

As an illustrative initial case we will examine Sierra Leone, because it shows evidence of both success and failure in approaching DDR. The process went through three distinct phases, each with a slightly different alignment of contributing actors, responsibilities and accountabilities. The phases were well-documented and reflect an ongoing dynamic of action, outcome and adjustment that captures lessons learned. For these reasons Sierra Leone represents an instance where collaboration would seem likely given the ongoing adjustment process and the opportunity to reflect on limitations in earlier iterations. It should be possible to analyse the causal connection between observed collaboration in this case and the phased iteration process leading up to it, or by the same token, the causal factors that inhibited collaboration even when the context seemed auspicious.

There is little consensus on the elements of successful DDR, though scholars have noted turf battles and trust factors as dynamics inhibiting success (see Muggah 2005; Weinstein and Humphreys 2005; Spear 2006; Knight and Ozerdem 2004). Our effort to focus specifically on collaboration, both why it occurs or fails to occur and its causal link to success, thus represents an important addition to the literature on both interorganisational networks and post-conflict intervention. We hope this analysis will provide the basis for building a guide to future successful implementation, that provides IGO and NGO personnel with tools for navigating the networking process.

Theoretical framework

Much of what we know about interorganisational relations stems from management theory, along with adaptations to the international dimension in the international organisation literature and a growing focus on NGOs in conflict resolution. Evan (1965) saw the problem as a systems theory involving networks of interacting organisations as constituent parts in what he labelled 'organizational sets'. Such sets often involve a 'focal organization' in relationships with cooperative bodies in what might be deemed 'input' or 'output' arrangements. The focal organisation serves as the lead while the cooperating actors function much like subcontractors. The relationship has also been described as establishing the terms of 'who drives', creating a clear leader/subordinate structure (Fawcett 2003). Evan writes:

> As a partial social system, a focal organization depends on input organizations for various types of resources: personnel, materiel, capital, legality and legitimacy, etc. Likewise the focal organization, in turn, produces a product or a service for a market, an audience, a client system, etc (1965, p. B221)

In contrast to well-defined organisational sets, Thayer (1981) favoured viewing the development of interorganisational networks in world politics along the lines of a 'structured non-hierarchy'. This entails inter-workings of units having 'ambiguous and fluid boundaries' which tend to 'penetrate and permeate each other' much as nation-states themselves have come to do in an 'interdependent' world (Jonsson 1986, p. 40). While organisational sets imply a system of focally and peripherally ordered units in a hub and spoke model, the structured non-hierarchy might have some elements of linear structure but in an *ad hoc* format that might change as needed from incident to incident. As the name implies, the non-hierarchy also has no necessarily leading or focal actor.

Thayer instead sees '"overlapping groups" at various levels, with some individuals [units] designated as "linking-pins" who act as communication channels through which consensus may be reached' (Jonsson 1986, p. 1). Instead of an authoritative actor anchoring the set, the structured non-hierarchy relies on certain organisations to mediate and build consensus among the group. The focal actor of the organisational set and linking pin actor of the structured non-hierarchy thus differ in their potential authority, role and leverage. Though the extent of formalisation can vary, two trends seem common: networks tend to be centralised around a group of organisations perceived as effective in generating collective action, and organisations try to reduce the risks of networking by partnering with their partners' partners (Atouba and Shumate 2010).

Within the field of international intervention several studies examine how integration worked or failed to work in a particular case, and identify general dynamics that would improve interaction. Ricigliano (2003) has proposed what he calls a Network of Effective Action (NEA) for organisations engaged in peacebuilding. The NEA would constitute a set of practices rather than structures, promoting integrated organisational response across the three areas of peacebuilding: political, social and structural. The principles that would guide the NEA connect closely to the four performance criteria for organisational networks identified by Nunez and Wilson-Grau (2003), which define equity, transparency, diversity and participation as key ingredients for successful interaction. Similarly, Garb and Nan (2006) endorse loose networks that allow for freedom among parties but encourage increasing coordination over time and shared engagement in the shape, process and direction of peacebuilding. On the flip side, factors that impede coordination include ambiguity of and lack of consensus on strategic objectives, the divide between national and international agendas and a failure to subsume logistic support to operational need as critical impediments to coordination (Last 1995, see also Brahimi 2000).

IGO networks

The fact that IGOs exist at three distinct levels in world politics – regional, extra-regional and global – suggests room for hierarchies, and indeed the UN Charter (Chapter VIII) set it up that way. The Charter places the Security Council at the top in authority, but with responsibilities potentially parcelled out at times to regional partners who need Security Council authorisation for their actions on specific topics (see Felicio 2009, pp. 13–14; Claude 1988). However, the divergence of organisational priorities and perspectives, the competing political agendas of major and minor organisational members and the obstacles cited above to IGO collaboration appear to fit very well with Thayer's concept of a structured non-hierarchy. No one organisation can be expected to focus on all questions, and since subunits, i.e. member states or organisational agencies and bureaus, tend to drive the organisations themselves, coalitions are likely to be *ad hoc*, even tentative and shifting before, during and after joint operations.[1] Regional involvement is also critical in sensitive areas like the Middle East, where neither the UN nor NATO can act fully effectively without regional approval, in spite of recognised capacity.

An additional factor likely producing and affecting structured non-hierarchies is the overlapping membership by some states in organisations at all three levels; states may be members of global, regional and extra-regional organisations, with different roles and interest preferences in each. Karen Mingst lays out the cross-cutting nature of IGO membership and notes that the multiple layers make coordination even more difficult, because a specific state's interest may accord with one IGO's agenda in a particular case but not another (Mingst 2010). Particularly in DDR, the target state as well as the remnants of separate militias are also central players, introducing yet another set of relationships, as when the various IGO member states also have independent relationships with the target state or rebels, which could either enhance or impede effective collaboration.

Muggah (2005, p. 244) has noted the varied functional and sectoral roles of certain IGOs in DDR specifically. Recognising the key linkage between sustained peace and successful development programmes, the World Bank has, for example, established criteria for post-conflict demobilisation and reintegration programmes, comprising some 27 or more projects in over 16 countries, 'often in close partnership with UN agencies who reluctantly handled disarmament, weapons destruction and reform of the security sector' (citing Colletta et al. 1996). Yet the World Bank itself was initially hampered by its restricted experience and purview in the disarmament sector, and indeed in the entire area of peacemaking and enforcement (as distinct from reintegration), which are viewed as UN and regional organisation functions (Bradley et al. 2002). Phases and segments of activity are therefore relevant in DDR, as some organisations might have expertise and capacity in one phase but not in others.

Coercive capacity is also a critical factor in DDR, and helps define which IGOs are most relevant at what times. While the UN and its associated agencies have superior international legitimacy, along with the advantage of sustained and reiterated experience and mission preparation, NATO and increasingly the European Union (EU), along with African IGOs, seem to have become the 'go-to' actors for enforcement, sanctions or robust military operations depending on geographic or political calculus. This positions the UN as the likely 'hub' organisation but also makes it clear that decentralised and flexible structures are likely to evolve in the DDR policy domain. Regional actors with local cultural sensitivity are increasingly seen to be crucial for legitimacy in particularly controversial cases. In West Africa, for example, the local connection of the Economic Community of West African States (ECOWAS) provided it with greater legitimacy than the UN in the eyes of the populace, in spite of the latter's greater capacity (Talentino 2005).

Further, in certain multilateral interventions, particularly those spilling over from initial peacemaking and peacekeeping to long term peacebuilding operations, the interorganisational set can become quite extensive, including the involvement of global organisations, regional IGOs and supportive functional agencies such as the World Bank, UN High Commission on Refugees (UNHCR), the International Red Cross and cooperating nongovernmental organisations (see Biermann and Siebenhuner 2009, p. 8). NGOs and functional agencies, though not directly involved in security issues, may have interests in the region that encourage them to lobby IGOs or otherwise try to influence responses and actions and can require difficult and frustrating interorganisational and inter-agency negotiation, harmonisation or division of labour (see for example Chandrasekaran 2012).

Jonsson (1986, Table 1) posits two sets of factors as fundamentally affecting the network pattern among collaborating organisations in any international mission: issue specific factors and organisation specific factors. Issue area and context relate to the combination of mission functions (e.g. security, reemployment, refugee relief) that must be performed. Implementation tasks are affected by whether they impact technically complex or more routine politics, whether there is issue polarisation in the IGO network, whether concentration or fragmentation of power exists among the IGOs and whether there is corresponding pluralist or authoritarian decision-making on the issue. Disarmament, for example, can be seen as a challenge unto itself, but can also be framed as a necessary adjunct of demobilisation and reintegration in the broader sense.

Organisation specific factors comprise such network related aspects as reachability, mobility, conspicuousness, constituents and leadership and determine the role an organisation plays in the network. A central or focal organisation would presumably direct the intervention goals and implementation, while a mediating organisation, the 'linking pin' role (Jonsson 1986), would negotiate varying visions with key actors to develop consensus while maintaining a presence, making its bureaucratic expertise available and applying leverage at key deliberations.[2]

One may posit that the variance in such factors brings certain organisations to the fore and helps account for the success or failure of missions in specific interventions. If collaboration more often follows organisational sets we should be able to identify a group of focal actors who are primarily responsible for the outcomes. If collaboration more often follows structured non-hierarchies, we should see both clear and shifting lines of responsibility and the emergence of certain organizations as essential consensus-builders.

Learning should thus be an important part of how well interorganisational networks function. Actors such as the UN, NATO, EU and AU may experience few or many operations together, may come to know better what to expect from each other and work harder to establish collaboration, and may become more inclined or disinclined to repeat certain processes, depending on their interpretation of past experience. Signs of this are evident, for example, in the emerging EU-AU partnership. This brings in the element of good versus bad news lessons or interpretations (constructivist perspectives) from prior involvements and initiatives. The fact that organisations seek partnerships with their partners' partners reinforces the good news concept, and suggests that trust and relationships are at the centre of effective collaboration (Atouba and Shumate 2010).

Smith and Schwegler (2010, pp. 284, 291ff) explore the trust issue and determine that it resides in the conditions necessary for effective organisational partnership. Presumably trust is facilitated by repeated positive or 'good news' experiences, such as defined successes in any of the three DDR functions, or repeated mutual enforcement efforts between those in the same organisational set. Repeated exposure and interaction yield benefits ranging from standardised operational procedures to common expectations and familiarity among personnel and staff (Garb and Nan 2006; see also Ricigliano 2003). Trust and predictability of response is easier the longer the organisations have been cooperating with each other, while a long history of discordant or perceived failed partnerships can have the opposite effect. Short term and *ad hoc* or unexpected collaborations presumably leave the most room for uncertainty, conjecture and suspicion, but the press of events in a constricted time frame might erase concerns about trust, at least for the interim, as parties move ahead in tandem out of practical necessity (Smith and Schwegler 2010, p. 305).

Disarmament and demobilisation

DDR by its very multifaceted nature is a good testing ground for the forms of interorganisational network analyses we have described. By embedding the process in what is often seen as 'phased' activity – moving from initial disarmament to selective weapon reductions and militia demobilisation to reestablishment of public order and governmental functions to economic reconstruction and development – inevitably a wide variety of organisations are brought into play, some with overlapping and some with very particular experiences, strengths and purviews.

At first glance, the UN may be viewed as an essential actor to DDR because of its centrality, legitimacy and potential ability to guarantee the terms of peace agreements to each of the formerly warring parties. But as Barker (2008, p. 102) notes, 'a significant challenge facing DDR programmes is the sheer number of actors involved in the process', as well as the fact that disarmament becomes embedded in larger peacebuilding challenges and strategies involving reconstituted governmental viability, rebuilding and promoting economic recovery and providing satisfactory regional or group security guarantees. The belligerent parties, other national parties from the government and/ or civil society, IGOs, regional organisations, NGOs and potentially other governments are also often heavily invested, and local armed groups can represent collaborators or spoilers, hoarding and trading key weapons even as other arms are confiscated or discarded.

Despite its appearance as a straight-forward set of functions, therefore, the DDR process can be significantly more complex than other, admittedly complicated peacekeeping domains, for example police reform, or rule of law, which are often handed over to one agency/entity to address or at least to serve as the primary focal point. Functional proliferation extends from the conception that DDR is multidimensional; though its primary outcome is 'security', it also relates to a variety of economic, social and political goals, as noted at the outset (see Schulhofer-Wohl and Sambanis 2010, p. 7). Indeed, while some groups will be the target of disarmament efforts, others will, of necessity, be armed for purposes of peacekeeping, law enforcement or public order. It is often difficult to distinguish former soldiers from insurgents and to know which groups should advisedly or legitimately retain arms out of political necessity. Particularly when 'transitional governments' are established for interim peacebuilding periods, varied factions are likely to play a role and to remain suspicious about their own security if their constituents are disarmed; certain IGO member governments, however, may remain suspicious if undesired groups remain armed.

Phase one of DDR is enforced rather than voluntary disarmament operations targeted at recent combatants, often administered by peacekeepers operating through 'organized, centralized, supervised, public' modalities (Faltas et al. 2001). Phase two activities come later in the process, are generally voluntary and

> are designed to reduce civilian arms possession. These 'phase two' interventions advance a combination of collective or individual incentives, are decentralized and often preceded by various penalties to deter illegal ownership. Examples of phase two interventions include 'weapons for development programmes', 'weapons lotteries', voluntary amnesties and 'weapons-free zones' [...]
>
> (Muggah 2005, p. 245)

These phases are not static, however, as new political agreements treat issues not covered in the original ceasefires and might necessitate tradeoffs of weapon confiscation in return for territorial or political concessions. This happened in Kosovo, for example, and delayed effective DDR. To be complete, micro-disarmament must be seen as part of a larger political agenda, but this agenda can, if poorly coordinated, pit IGOs and their members against each other.

Mission definition and coordination is always a challenge in multilateral peacemaking operations, as noted by the UN Brahimi Report on joint interventions, which warned of sending armed forces into ill-defined and ill-conceived operations. Gender bias and group or regional differences can also complicate overall mission attainment. DDR seems to offer a clear mission but is subject to such challenges as determining success criteria and programme labelling. Communications are sometimes vague and inadequate, in the past often focusing, for example, on numbers of confiscated weapons rather than on sustained peace, political reform or reconstruction. Several analyses of specific or multiple cases highlight lack of strategic consensus as an inhibitor for organisational coordination (Last 1995; Last and Vought 1994; Hayes and Weatley 1996). In addition, disarmament and reintegration often carry financial incentives that have led, in some instances, to greater conflict as groups react in frustration at being left out of the benefits. Financial incentives can also encourage weapon black markets to proliferate in the midst of economic chaos (Faltas et al. 2001; Muggah 2005, p. 247).

The implementation of DDR programmes can also be challenged by regional developments, as in Liberia, where deployment of simultaneous disarmament programmes in Sierra Leone and Côte d'Ivoire and the lack of communication and coordination between UN officials and their counterparts implementing these programmes encouraged demobilising groups to seek the highest bidder through black market and cross-border activity. The announcement that ex-combatants in Côte d'Ivoire under the DDR program would be receiving more money than Liberians created potential for a cross-border market and weapon smuggling as 'many Liberian fighters will be holding back weapons in order to cash them in next door' (Nichols 2005, p. 128, see also Paes 2005). In addition to lack of communication among DDR programmes as a confounding factor that jeopardised implementation in Liberia, the 'quick fix' and uncritical 'one size fits all' approach (Jaye 2009, p. 13; Conference Report 2005) to replicating the programme for Sierra Leone without any critical consideration of the conditions on the ground impaired the effectiveness of the programme. The success of DDR is thus impacted by exogenous and endogenous factors that shape how organisational coordination takes place.

Case study: Sierra Leone

The factors defined as relevant to both organisational collaboration and network patterns are evident in Sierra Leone. Sierra Leone's DDR process was implemented in three distinct phases and is generally judged a success (Solomon and Ginifer 2008). Throughout the three DDR phases the network of organisations evolved through a set of practices that were not clearly evident at the outset but developed through repeated interactions, trust-building via transparency and participation and connection of logistical support to operational need (see Nunez and Wilson-Grau 2003; Last 1995). The network pattern also evolved by more effectively matching the organisation specific factors with the issue specific needs over time (see Jonsson 1986), as explained below.

Reintegration was the weakest link in the chain, which likely contributed to regional insecurity dynamics. Former fighters were not effectively incorporated back into society, allowing the perpetuation of a combatant culture, and one observer stated:

> The Reintegration phase was also hampered by the ex-combatant hesitation to return to their communities for fear of reprisals for their actions during the conflict and also, these communities did not have the capacity to absorb all these returnees. The outcome was the migration of these ex-combatants to other West African countries, e.g. Liberia and Cote d'Ivoire to sell their labor as mercenaries and perpetuate instability...
>
> (African Development Bank, 2011, p. 6)

Further, the process unfolded in a halting way, which allowed learning across phases but also created a dynamic of implementation and relapse that meant, in some cases, that the same individuals were disarmed more than once.

All phases of the formal DDR program were designed and overseen by the Government of Sierra Leone working through the National Committee for Disarmament, Demobilisation, and Reintegration (NCDDR). This was a distinctive aspect of the Sierra Leone case because it gave formal ownership to local authorities even though they were dependent on international actors for implementation. The Committee was judged by observers as 'one of the few functional domestic institutions', though it was heavily controlled by donors (Meyer 2007). In the first two phases the NCDDR served as a hub to the organisational set, but by the third phase the network pattern shifted more toward a non-structured hierarchy as subunits of the NCDDR began to play greater roles. The proliferation of interests and donors reflects a common dynamic of DDR processes, where the complexity of tasks opens channels for many competing voices. To address the international/local tension the NCDDR established weekly coordination meetings and metrics for planning, thereby establishing itself as a focal organisation.

Although it did not work particularly well in phase one, this approach also established the set of practices – participation, communication and at least nominal transparency – that is identified as enabling effective collaboration (Ricigliano 2003). Some of the obstacles identified by Last (1995) were also present, however, as the capacity needed for logistical support was not correctly assessed beforehand, and national and international agendas were not clearly aligned. The ECOWAS Monitoring Group (ECOMOG) for example, was not fully able to provide the security that was essential for disarmament and engage in actual disarmament and weapons collection, and as an observer mission, the United Nations Observer Mission in Sierra Leone (UNOMSIL) could provide little support (Tesfamichael et al. 2004). Further, reintegration programmes in particular were diverse in terms of how closely they targeted ex-combatants versus overall community integration, and how closely linked they were to disarmament and demobilisation versus social capacity building (Meyer 2007; see also Pugel 2006).

The role played by both ECOMOG and UNOMSIL changed entirely in phase two, when the United Nations Mission in Sierra Leone (UNAMSIL) took over after ECOMOG's phased withdrawal.[3] That might seem to augur a potential focal role for the UN, but UNAMSIL was inadequately funded and equipped for the role assigned. It could not manage disarmament, and required assistance from Britain's Emergency Response Team (ERT). Prominent IGOs were thus limited by function in both phase one and two. Neither ECOMOG nor UNAMSIL could play both a security and DDR role – which reinforced the role of local actors even as it exposed the lack of implementation capacity across the network. Britain's ERT also took on what might be described as a linking pin role, because it became a primary facilitator for moving other actors toward a point of effective capacity. Although the organisational network seems clear in this phase, some elements of a non-structured hierarchy were also present due to the UN's limited abilities. Aligning implementation and goals proved a major problem among organisations, as disarmament started before demobilisation centres were ready and certain weapons used by the civil defence forces were exempted, creating confusion about who should do what and when. Given UNAMSIL's inability to disable weaponry, the World Bank stepped in to finance consultants to assist the NCDDR in developing this capacity (Tesfamichael et al. 2004).

Phase two ground to a halt because of these problems, though the nearly year-long period between phases two and three marked a notable period of IGO cooperation. The ERT, which had been largely responsible for running the disarmament camps, was withdrawn in 2000. That left a significant void in management, but also opened the door for other organisations to transition their roles to focus on tasks that fit more closely with their comparative advantage. UNAMSIL, the World Food Programme, the International Medical Programme and local contractors all stepped in to take on various tasks, allowing the government to keep the camps running. UNAMSIL's role in particular increased significantly, as it took on greater levels of camp management, an area where it did have sufficient capacity and could relieve some of the coordination responsibility from the NCDDR. Although the ERT had served a broad role, providing overall support for the process and security throughout the country and building UNAMSIL's operational capacity, after it left these tasks were divided and parcelled out in a way that enhanced the non-structured hierarchy and allowed both greater specialisation and leadership in each specific area. World Bank funding also gave the NCDDR some implementation capacity, which reinforced its role.

In the interim between phases two and three the government focused on reintegration and lessons learned. One key change was to reduce the role of the NCDDR, which was hampered by Revolutionary United Front (RUF) representation. Although their involvement was certainly seen as critical, RUF members had little confidence in, or commitment to the overall peace process. By shifting greater responsibility to the Executive Secretariat and Technical Coordinating Committees (TCCs) of the NCDDR, the government was able to focus on implementation tasks rather than broader policy considerations. The central organisation thus changed from the NCDDR itself to a subunit of the Committee. The Executive Secretariat improved effectiveness in DDR through a variety of means, including honesty and openness, which encouraged partner input and allowed more effective coordination of operations with other organisations (see Morse and Knight 2002; Jennings 2007). Tesfamichael et al. (2004) argue that: 'By and large, the institution came to gain the confidence not only of the major actors but also of the individual ex-combatants it was set up to serve'. They also suggest that the availability of informal back channels played a significant role in the process, with a variety of informal means used to build confidence among stakeholders in the process and maximise interactions among organisations. This is a factor that Garb and Nan (2006) cite as critical in creating effective systems of coordination.

It also seems clear that the adaptation of organisations on the ground played a critical role in coordination and overall effectiveness. UNAMSIL expanded the writ of UN support and played a central role in disarmament in phases two and three. The role initially defined for it fell outside its organisational capacities, but the departure of the ERT opened up logistical and management needs in running the camps that proved a better fit for UNAMSIL's abilities. The shift therefore occurred partly through institutional channels and the vacuum left for administration of the camps, and partly through individual initiative. The UNAMSIL Force Commander took on what can be described as a linking pin role after 2000, defined by relationship development among the various fighting groups, particularly the RUF, which had acted as a spoiler in the earlier phase. Other organisations contributed by taking on smaller, more specific parts of the process allowing greater concentration of focus. For example, the World Food Programme was able to step in on feeding of ex-combatants, taking on that aspect of camp management, while UNICEF spearheaded reintegration of child soldiers (see Tesfamichael et al. 2004; Gberie 2006). This seems to have had positive effects for organisations, which could focus on more narrowly defined tasks that fit more directly into their area of expertise, and the process, which benefited from greater capacity and authority applied to its constituent parts.

The World Bank's role in providing funding and technical support was crucial throughout DDR, particularly in terms of allowing the Sierra Leone government to remain the focal organisation for the overall process. Without the technical expertise and funds provided by the Bank, the NCDDR and Secretariat's ability to establish overall authority over the process would have been much diminished. Another factor contributing to successful coordination was the creation of structures within structures, which improved transparency and reduced political conflict. In addition, the Executive Secretariat of the NCDDR structures, such as the Multi-Donor Trust Fund (MDTF) and Financial Management and Procurement Unit (FMPU), which emerged after the reorganisation in 2000 and the period of reflection and careful planning between phases two and three, allowed focus on smaller, very specific issues, shielding part of the process from the mismanagement or corruption that might have been seen in other cases, and helped confidence building among parties (Tesfamichael et al. 2004; United Nations 2010).

This case therefore reveals important considerations for further study. First, organisational network and non-hierarchical models are unlikely to be static over time. The Sierra Leone process started closer to the organisational network model and evolved toward a non-hierarchical structure in the third phase, leading to overall success by responding to contingencies. Further, even in the organisational network the theory does not assume full capacity by any organisation. The NCDDR played a focal role by outlining the process and serving as the hub through which other organisations interacted in phases one and two, but other organisations played critical roles as well, either in direct action or facilitation of others' action. Second, organisational learning certainly characterises this case. At some points the learning came from intentional group reflection, while in other cases it came from organisations filling areas of action by default. In either case, at the conclusion of the process it is easy to see evidence of redirection and reorganisation throughout the process.

Finally, three of the four criteria defined by Nunez and Wilson-Grau (2003) for effective collaboration make an appearance here, which likely contributed to the outcome, though we have not fully defined a causal connection from collaboration to effectiveness. Transparency, diversity (in terms of both local and international involvement), and participation (having a say in the process) were present to at least some degree, particularly in phase three. Specific actors rose to important mediating and facilitating roles allowing organisations, whether or not full equity was present, to be consciously cultivated and responsibly included over time.

Conclusion

Due to technical and financial constraints, no actor can carry out a DDR program without cooperation with other stakeholders; thus, coordination and integration are a *sine quo non* in these programmes. As Sierra Leone indicates, it is critical to establish structures that are adaptive and capable of designing, planning and implementing DDR in an integrated manner by taking other possible and current partners into account. Lack of coordination among the partners hampers the effectiveness of DDR programmes. Hence, to be truly effective, disarmament and demobilisation programmes should be in communication and consultation with appropriate parties, including the United Nations, regional, sub-regional organisations, national and local stakeholders. The UN peacekeeping operations apparatus is seen as a natural focal point for such networking, but our initial review shows that the relationships are essentially and structurally non-hierarchical across phases. This suggests a hypothesis for further study – the greater the presence of the four criteria of equity, transparency, diversity and participation, the greater the likelihood of effective collaboration and mission success.

The role an authoritative focal organisation can play in the interorganisational set is to manage the collaboration, particularly in the early stages or phases and to parcel out and coordinate the early division of labour to hinder duplication of efforts. This coordinating agency, therefore, can promote if not ensure unity and coherence in policy implementation and strategy and help the programme to get back on track when things go wrong, diminishing finger pointing among participating organisations and critical governments. But the Sierra Leone case also highlights the limitations focal organisations may have, particularly if the division of labour they establish is ineffective or if they are beset by 'spoilers'. In this case three divergent views of ownership were at work throughout the DDR process, planning as ownership (the NCDDR), funding as ownership (World Bank), and doing as ownership (the ERT and ECOMOG in the early stages). The UN initially fell into no category and had to re-envision its role in concert with others. Ideally, the central organisation would be trusted and backed by collaborating IGOs, NGOs and national stakeholders, thus establishing a sense of ownership among local and national actors. Most observers rated the NCDDR as effective and open, which helped it build trust between the implementers and ex-combatants, and the Executive Secretariat's impact, combined with the UNAMSIL force commander in phase two, prevented the RUF from 'turning into spoilers of the peace process' (Banholzer 2014, p. 22). Just as important, it must have a firm command of other organisations' capacities or it will quickly lose its grip on the process.

This chapter only scratched the surface in outlining interorganisational networking models in accounting for the effectiveness of DDR and other post-conflict programmes. The diverse tasks within DDR and the effort to incorporate local ownership will mean that, inevitably, and even when processes follow best practices, no single entity will be able to claim leadership on or carry out all aspects of the task. In Sierra Leone, the DDR process was further complicated by the fact that the roles envisioned for IGOs did not always match with their abilities, highlighting the importance of collaborative mission definition from the outset. This raises another topic for further study – the frequency with which and steps taken to ensure role and capacity align. DDR processes require comprehensive and objective assessments of capabilities and right-fit for each participating entity, though in this case the right-fit frequently evolved by default. Further research, and particularly interviews with agencies that have participated in joint DDR operations, should focus on systematic cross-sectional studies examining the extent to which current practices and interorganisational networking contribute to or hamper the success of these programmes.

Notes

1 Non-hierarchies often even emerge *within* organisations such as the UN, which may have many different and practically autonomous units working within any one geographical region. Muggah's (2005) description of turf battles applies to UN agencies, which are technically part of a single IGO but functionally tend to operate independently of each other.
2 See Garb and Nan (2006) on the Abkhaz-Georgia negotiations, though with primary focus on NGO networks.
3 ECOMOG withdrew because it could not sustain the force due to the pending withdrawal of Nigeria, its largest contributor, as a result of democratic elections in that country in 1999 and the transition from military to civilian rule. See Doyle 1999.

References

African Development Bank. 2011, *The Role of Disarmament, Demobilization and Reintegration Programs in Post-Conflict Reconstruction: Some Lessons Learnt*. Abidjan: African Development Bank.

Atouba, Y. and Shumate, M. 2010, 'Interorganizational Networking Patterns Among Development Organizations', *Journal of Communications*, 60(2): 293–317.

Banholzer, L. 2014, *When Do Disarmament, Demobilization and Reintegration Programmes Succeed?* Bonn: German Development Institute.

Barker, D. 2008, 'Towards a More Comprehensive DDR', *Journal of Politics and International Affairs*, 4: 101–114.

Biermann, F. and Siebenhuner, B. (eds). 2009, *Managers of Global Change: The Influence of International Environmental Bureaucracies*. Cambridge: MIT Press.

Bradley, S., Maughan, P. and Fusato, M. 2002, 'Sierra Leone: Disarmament, Demobilization, and Reintegration (DDR)', Africa Regional Findings and Good Practice Infobriefs; No. 81, viewed 1 May 2017, https://openknowledge.worldbank.org/handle/10986/9754.

Brahimi, L. 2000, 'Report of the Panel on United Nations Peace Operations', A/55/305-S/2000/809, viewed 1 May 2017, www.un.org/documents/ga/docs/55/a55305.pdf.

Cimbala, S. J. and Forster, P. K. 2010, *Multinational Military Intervention: Nato Policy, Strategy and Burden Sharing*. Burlington, VT: Ashgate Publications.

Chandrasekaran, R. 2012, *Little America: The War within the War in Afghanistan*. New York, NY: Alfred A. Knopf.

Colletta, Nat J., Kostner, M., and Wiederhofer, I. (1996), 'Case Studies in War-to-Peace Transition: The Demobilization and Integration of Ex-combatants in Ethiopia, Namibia, and Uganda'. World Bank, viewed 1 May 2017, http://documents.worldbank.org/curated/en/385411468757824135/Case-studies-in-war-to-peace-transition-the-demobilization-and-reintegration-of-ex-combatants-in-Ethiopia-Namibia-and-Uganda.

Conference Report. 2005, *Disarmament, Demobilization, Reintegration (DDR) and Stability in Africa*. Freetown: United Nations, Freetown, 21–23 June.

Claude, I. 1988, *States and the Global System: Politics, Law, Organization*. London: MacMillan Press.

Doyle, M. 1999, 'Nigerian Election "Threatens" Sierra Leone', *BBC News*, February 27, viewed 1 May 2017, http://news.bbc.co.uk/2/hi/africa/287236.stm.

Evan, W. M. 1965, 'Toward a Theory of Inter-Organizational Relations', *Management Science*, 11(10): 217–230.

Faltas, S., McDonald, G., Waszinck, C. 2001, 'Removing Small Arms from Society: A Review of Weapons Collection and Destruction Programmes', Occasional Paper, No. 2. Geneva: The Small Arms Survey, viewed 1 May 2017, www.smallarmssurvey.org/fileadmin/docs/B-Occasional-papers/SAS-OP02-Weapons-Collection.pdf.

Fawcett, L. 2003, 'The Evolving Architecture of Regionalization', in M. Pugh and W. P. S. Sidhu (eds), *The United Nations and Regional Security*. Boulder, CO: Lynne Rienner, 11–30.

Felicio, T. 2009, 'The United Nations and Regional Organizations: The Need for Clarification and Cooperation', In J. Koops (ed), *Military Crisis Management: The Challenge of Interorganizationalism*. Brussels: Egmont, 13–20.

Garb, P. and Nan S. A. 2006, 'Negotiating in a Coordination Network of Citizen Peacebuilding Initiatives in the Georgian-Abkhaz Peace Process', *International Negotiation*, 11(1): 7–35.

Gberie, L. 2006, 'Sierra Leone: Remembering a Difficult Disarmament Process', *Pambazuka News*, September 28, viewed 1 May 2017, www.pambazuka.org/global-south/sierra-leone-remembering-difficult-disarmament-process.

Hayes, M. D. and Weatley, G. F. 1996, *Interagency and Political-Military Dimensions of Peace Operations: Haiti, A Case Study*. Washington, DC: National Defense University.

Jaye, T. 2009, *The Transitional Justice and DDR: The Case of Liberia*. New York, NY: International Center for Transitional Justice.

Jennings, K. M. (2007), 'The Struggle to Satisfy: DDR through the Eyes of Ex-Combatants in Liberia', *International Peacekeeping*, 14(2): 204–218.

Jonsson, C. 1986, 'Interorganization Theory and International Organization', *International Studies Quarterly*, 30(1):39–57.

Knight, M. and Ozerdem, A. 2004, 'Guns, Camps and Cash: Disarmament, Demobilization and Reinsertion of Former Combatants in Transitions from War to Peace', *Journal of Peace Research*, 41(4): 499–516.

Last, D. M., Vought, D. 1994, *Interagency Cooperation in Peace Operations*. Conference Report. Fort Leavenworth, KS: US Army Command and General Staff College.

Last, D. 1995, 'Peacekeeping Doctrine and Conflict Resolution Techniques', *Armed Forces and Society*, 22(1): 187–210.

Liebler, C. and Ferri, M. 2004, *NGO Networks: Building Capacity in a Changing World.* Washington, DC: USAID, Office of Private and Voluntary Cooperation.

Meyer, S. 2007, 'Sierra Leone: Reconstructing A Patrimonial State', *Fride*, viewed 1 May 2017, http://fride.org/descarga/BGR_SierrLeo_ENG_may07.pdf.

Mingst, K. 2010, *International Organizations: The Power and Process of Global Governance.* 2nd ed. Boulder, CO: Lynne Rienner.

Morse, T., and Knight, M. 2002, *Lessons Learned from Sierra Leone Disarmament and Demobilization Programme Assessment Report.* Freetown: Government of Sierra Leone and the World Bank.

Muggah, R. 2005, 'No Magic Bullet: A Critical Perpective on Disarmament, Demobilization and Reintegration (DDR) and Weapons Reduction in Post-Conflict Contexts', *The Commonwealth Journal of International Affairs*, 94(379): 239–252.

Nichols, R. 2005, 'Disarming Liberia: Progress and Pitfalls'. The Small Arms Survey, viewed 1 May 2017, http://www.smallarmssurvey.org/fileadmin/docs/D-Book-series/book-01-Armed-and-Aimless/SAS-Armed-Aimless-Part-1-Chapter-04.pdf.

Nunez, M. and Wilson-Grau, R. 2003, 'Towards a Conceptual Framework for Evaluating International Social Change Networks', viewed 1 May 2017, www.mande.co.uk/docs/Towards%20a%20Conceptual%20Framework%20for%20Evaluating%20Networks.pdf.

Paes, W. 2005, 'The Challenges of Disarmament, Demobilization and Reintegration in Liberia', *International Peacekeeping*, 12(2): 253–261.

Pugel, J. 2006, *What Fighters Say: A Survey of Ex-Combatants in Liberia.* New York, NY: United Nations Development Programme.

Ricigliano, R. 2003, 'Networks of Effective Action: Implementing an Integrated Approach to Peacebuilding', *Security Dialogue*, 34(4): 445–462.

Schulhofer-Wohl, J. and Sambanis, N. 2010, *Disarmament, Demobilization, and Reintegration Programs: An Assessment.* Sweden: Folke Bernadotte Academy.

Smith, L.R. and Schwegler, U. 2010, The Role of Trust in International Crisis Areas: A Comparison of German and US-American NGO Partnership Strategies', In D. Skinner, M. Saunders, G. Dietz, N. Gillespie, R.J. Lewicki (eds), *Organizational Trust: A Cultural Perspective.* Cambridge: Cambridge University Press, 281–310.

Solomon, C. and Ginifer, J. 2008, 'Disarmament, Demobilisation, and Reintegration in Sierra Leone', Centre for International Cooperation and Security, University of Bradford, viewed 1 May 2017, www.operationspaix.net/DATA/DOCUMENT/4024~v~Disarmament_Demobilisation_and_Reintegration_in_Sierra_Leone.pdf.

Spear, J. 2006, 'From Political Economies of War to Political Economies of Peace: The Contribution of DDR after War Predation', *Contemporary Security Policy*, 27(1): 168–189.

Talentino, A. K. 2005, *Military Intervention after the Cold War: The Evolution of Theory and Practice.* Athens: Ohio University Press.

Tesfamichael, G., Ball, N., Nenon, J. 2004, *Peace in Sierra Leone: Evaluating the Disarmament, Demobilization, and Reintegration Process.* Washington, DC: Creative Associates International.

Thayer, F.C. 1981, 'Organization Theory, Political Theory, and the International Arena: Some Hope but Very Little Time', paper presented at the International Studies Association annual conference, Philadelphia, PA, 18–21 March, 1981.

United Nations. 2010, *DDR in Peace Operations: A Retrospective*. New York, NY: United Nations Department of Peacekeeping Operations.

United Nations. 2016, 'United Nations Peacekeeping Webpage', viewed 1 May 2017, www.un.org/en/peacekeeping/issues/ddr.shtml.

Weinstein, J. M. and Humphreys, M. 2005, 'Disentangling the Determinants of Successful Disarmament, Demobilization, and Reintegration', Working Paper No. 69, Center for Global Development, viewed 1 May 2017, www.cgdev.org/sites/default/files/4155_file_WP_69_0.pdf.

9 From MK-Ultra project to Human Terrain System

Militarisation of social sciences – ethical dilemmas and future prospects

Michał Pawiński

Introduction

The importance of warfare in human affairs has been the focus of intense research in a wide range of disciplines from biology to history, psychology and political science. Each discipline adds a different dimension to our understanding of the causes and role of warfare in the past, present and future. The outset of the 21st century did not bring the end of history, but rather a clash of civilisations marked by the events of 11 September 2001. Traditional methods of warfare related to the revolution in military affairs have proven to be inadequate to counter threats presented by contemporary non-state actors. Today's conflicts are increasingly waged among the people instead of around the people. It is therefore not a surprise that social scientists are recruited by the military institutions.

However, history tells us that when academia mingles with war, effects are not always beneficial for the subjects of study. The past projects, like MK-Ultra, that used and funded anthropologists, psychologists and behavioural scientists to study the effectiveness of mind control, brainwashing and interrogation and torture techniques, reminds us that the social sciences' involvement in military operations and intelligence collection might not be ethically and morally acceptable. This chapter will bring past experiences to evaluate contemporary ethical dilemmas related to the Human Terrain System that once again found intelligence agencies seeking help from academia. The chapter proceeds in the following manner. Part one will briefly discuss the historical development of cooperation between academia and governmental, military and intelligence agencies (GMI). Part two will elaborate on the reasons behind the waning collaboration between social scientists and GMI in the 1970s. Part three enters the debate about ethical dilemmas associated with the Human Terrain System. The last part proposes three future possible models of collaboration between social scientists and GMI: at the academic and think tank levels; at the strategic level and during the political decision-making process; and at the operational and tactical levels. It is argued in this chapter that the social sciences and social scientists can contribute to the reduction of pain and suffering of people during wars, however, it is necessary to understand the limits and boundaries of such engagement and possible abuses of such cooperation.

From behavioural studies to mind control projects

During World War II the Japanese and the American military cultures differed sharply from each other. Traditional code of the warrior, *bushido*, required the Japanese soldiers to treat their own life with disregard, despise surrender and embrace suicide. The most important principle in the martial code stood loyalty to the emperor. The honour and will to fight by the Japanese had been proven throughout the last year of the war. On the night of 9–10 March 1945, over 344 B-29 bombers carried and released 2,000 tons of incendiaries against Tokyo, causing devastation and a terrific inferno, killing 83,000 men, women and children. This did not bring about the capitulation, neither did the Little Boy, a nuclear bomb with an energy of approximately 15 kilotons of TNT that was dropped on Hiroshima on 6 August 1945. Only after the second atomic nightmare that was created on 9 August 1945 in Nagasaki did the Japanese accept the Potsdam Declaration (Lynn 2003, pp. 246–280). It is therefore hardly surprising that senior US military leaders and President Franklin D. Roosevelt perceived the Japanese as culturally incapable of surrender. Hence, the Bureau of Overseas Intelligence of the Office of War Information was ordered to study the view of the emperor in Japanese society. The two most important reports were written by anthropologists. The first one, produced by Geoffrey Gorer (1943), *Themes in Japanese Culture*; the second by Ruth Benedict (1946), *The Chrysanthemum and the Sword*. Both recommended to the Office of War Information to keep the Japanese emperor intact, as this would facilitate post-war relations and the rebuilding process. Whether the reports had been read by the highest authorities remains unknown. The fact is that, after the conclusion of war, the US government decided to act upon their recommendations. Many scholars perceived both reports as a new kind of applied anthropology, and as a beginning for the transdisciplinary study of national character. It became imperative that any conflict requires knowledge of the enemy's society, and knowledge of the allies to be able to fight united (McFate 2006).

The Cold War period happened to be fertile ground for further studies and experiments. Most of the social scientists involved in the making of the Cold War enemy were behavioural scientists, an interdisciplinary academic coalition for addressing diverse social and political concerns around the world. In contrast to the existing traditional approaches in the social sciences from the 1870s to the 1920s (Cravens 2012, pp. 117–137), the behavioural sciences championed classic liberal notions of autonomous individuals who belonged to one or another social group. In short, the 'I' and 'We' were tightly fused together, one being a manifestation and representation of the other, and vice versa. Society was considered to be a system of systems, an organic whole, consisting of individual nods. Therefore, social scientists could begin their analysis with either individual or the group. Such views were in opposition to the processes shaping the ideological struggle between the capitalist and communist world. Behavioural scientists discounted the power of ideas and values as motivating forces and treated ideology and belief systems as rationalisations of behavioural modes that could be explained through examining individuals. At that time, the social sciences depended heavily on a patronage system including military, propaganda and intelligence agencies, as well as influential private institutions, including the Carnegie Corporation, Rockefeller Foundation and Ford Foundation. This coalition of psychologists, sociologists and political scientists armed with intellectual weapons aimed at controlling and predicting the behaviour of friend and foe, got their chance to prove their assumptions during and after the Korean War (Solovey 2012, pp. 1–25).

Once the protracted nature of the Korean War had sunk in, the US army established the Office of the Chief of Psychological Warfare. The programme employed academic advisors who were associated with the RAND Corporation and two army's major think tanks – the Operations Research Office and the Human Relations Research Office. Academic scholars were responsible for designing propaganda leaflets that later were disseminated above the heads of the enemy (Chung 2004). Within the first 18 months of the Korean conflict, American aircrafts and artillery disgorged a billion of such fliers. However, when it came to the peace negotiations the focus of behviouralists shifted from civilians to leadership. Harold Lasswell, an influential American political scientist of the Cold War, argued that the 'government is always government by the few, whether in the name of the few, the one, or the many' (Dye et al. 2011, p. 1).

The most conspicuous attempt to fuse psychoculture and elite studies was associated with the work of Nathan Leites. In *The Operational Code of Politburo* and *A Study of Bolshevism* Leites pointed out that ideology had little importance for political decision-makers. Hence, weaponising capitalism against communism in an ideological struggle would be a futile endeavour. The key to undermining communism was through pinpointing psychological vulnerabilities in the 'Bolshevik mind'. Leites's operational code established an analytical framework to map psychological perceptions of the external and internal world of the Soviet elite. He further claimed that the code could be applied for deciphering all major strains of communist leadership, including its Asian varieties (Leites 1951). Hence, his publication had been a primary, if not the only, guide used by American negotiators in their efforts to decode their enemy interlocutors during Korean War peace talks. According to Herbert Goldhamer, who introduced operational code into negotiations, communist strategy was neither determined by a realistic assessment of circumstances, nor by historical cultural code. Their strategy was derived exclusively from a canon of political writings and inflexible behavioural modes. However, due to the lack of social scientific support, combined with limited understanding of operational code by the delegates, and given the very tedious, and prolonged nature of the actual negotiations, out of all the rules of communist behaviour, American negotiators appeared to rank what Leites called 'the calculus of the general line' as the essence of the code. It led to a superficial description of the enemy into a good-bad division, and conviction to never concede anything to the communists merely to make progress (Robin 2001, pp. 124–143). The focus on the mind had inadvertent consequences that crossed the ethical Rubicon of further academic involvement with governmental, military and intelligence institutions.

The political, social and intellectual atmosphere of the Korean War period reflected concern and fascination with the 'enemy within'. The fears that the enemy could take control over the mind of a soldier, agent or any civilian and use them as a double-sword weapon, resulted in commencement of secret brainwashing and mind control endeavours by the CIA. In the early 1970s, thousands of pages of the US government's classified documents were released. They contained information about CIA's covert programme known as MK-Ultra. During a Joint Hearing before the Select Committee on Intelligence, Admiral Stansfield Turner, CIA director between 1977–1981, asserted that 'we had two CIA prisoners in China and one in the Soviet Union, and we were concerned as to what kinds of things might be done to them' (Senate Select Committee on Intelligence 1977, p. 43). To counter those fears, it was necessary to design a project with a simple aim: to gain, through behavioural control, valuable information from the opponent. For that purpose, the MK-Ultra engaged over 80 universities, think tanks and other institutions, and more than 180 non-governmental scholars, who mostly unwittingly had been working on 149 subprojects to study methodically whether the effective forms of mind control, brainwashing, interrogation and torture techniques could be achieved. For instance, psychiatrist Raymond Prince received funding to undertake 'transcultural psychological studies' in Nigeria. The CIA's view was that his study

> will add somewhat to our understanding of native Yoruba psychiatry including the use of drugs, many of which are unknown by Western practitioners. It also will assist in the identification of promising young (deleted by CIA censors), who may be of direct interest to the Agency (Price 2007, pp. 8–13).

Table 9.1 lists some of these projects.

From Camelot to Baghdad

Testifying before the US Congress in 1965, Robert L. Sproull, then director of Defense Advanced Research Projects Agency, said:

> it is our primary thesis that remote area warfare is controlled in a major way by the environment in which the warfare occurs, by the sociological and anthropological characteristics of the people involved in the war, and by nature of the conflict itself' (McFate 2006, p. 35).

Table 9.1 Grants funded by the CIA, 1960–1963

Grant	Researcher	Field	Grant Size
Comparative study of Chinese personality	William G. Rodd	N/A	$3,000 US
Aspects of upper class culture among the international elite of Japan	Leon Stover	Anthropology	$3,000 US
Emerging socio-political roles of scientists and managers in the USSR	Albert Parry	Russian studies	$5,000 US
Changing patterns in the Chinese family	Lucy J. Huang	Sociology	$5,775 US
Pattern recognition	W.W. Bledsoe	Psychology	$45,000 US

Source: Price (2007), p. 11.

The 1964 Project Camelot was designed to determine the feasibility of developing a general social system model which would make it possible to predict and influence politically significant aspects of social change in the developing nations. Scholars were to find procedures for assessing the potential for internal war within society and identify actions that a government might take to mitigate those conditions. On the list of countries selected for study were mainly Latin American states, like Bolivia, Colombia and Paraguay. However, from the very beginning Project Camelot was ill-conceived, inappropriately funded and poorly coordinated. The information about the project leaked to the vice chancellor of the University of Chile, then to the Chilean Senate, and subsequently to the national press of Chile. The rage and anger spread among Chilean society, and diplomatic protests were raised. Operation Camelot was characterised as 'intervention', 'imperialism' and 'a vast continental spy plan' by the United States. The memory of disastrous Bay of Pigs invasion in Cuba (1961), and the US military intervention in the Dominican Republic in 1965, further exaggerated the situation. Soon after its inception, on 8 July 1965, the project was put to an end (Lucas 2009, pp. 56–63).

Although political controversies were a significant factor in abandoning Camelot, academic disagreements and incapability to find a common ground demonstrated the weaknesses of integrated behavioural sciences, and centrally organised, government funded projects. Anthropologists, psychologists, sociologists and political scientists become restless in the presentation of different approaches to the problems, therefore they could not agree on a binding theoretical framework. While some scholars supported rational choice theory, others anticipated irrationality in collective behaviour; where one assumed that collective behaviour was the aggregate sum of individual actions, the other argued for fundamental differences between emotional, exaggerated behaviour of the crowd and the normative behaviour of individuals. There was even no agreement on the definition of social change. Anatol Rapoport, a biologist from the University of Michigan, argued that the search for a unified language of science was counterproductive if not futile. According to his view 'the "total" society is far too complex to be encompassed by a single set of concepts' (Robin 2001, pp. 206–225). Political repercussions, combined with academic standstills and ethical dilemmas, made any future similar projects prone to fiasco. Additionally, controversies surrounding the Thailand Study Group in the early 1970s (Hinton 2002), along with CORDS and the Phoenix programme during the Vietnam War (Andrade and Willbanks 2006), further exacerbated problems mentioned in Project Camelot and MK-Ultra. Therefore, the end of the Vietnam War and subsequent revolution in military affairs brought about disengagement of social scientists from GMI sponsored endeavours.

The American military strategy of the 1990s was dominated by technology. Revolution in military affairs envisaged a future in which modern weapon systems, precision munition, rapid mobility and nearly instant access to information at strategic, operational and tactical levels of warfare, would give dominant, empowering advantage over any type of enemy. The first Gulf War had proven the point. Through 'shock and awe' the victory was achieved in just 6 months, 3 weeks and 5 days since the outset of the conflict (Ulleman et al. 1996). Moving forward to the year 2001, Operation 'Enduring Freedom' commenced on 7 October 2001 with an aerial bombing campaign. In the first few months, the Taliban were ousted from power in Afghanistan and fragmented as a political actor. 'Enduring Freedom' was not intended or designed to be a stabilisation operation, it was a pure exemplification of the Rapid Dominance doctrine (Conetta 2002, pp. 4–7). 20 March 2003 marked the beginning of Operation 'Iraqi Freedom'. It brought large-scale 'shock-and-awe' attacks on Iraqi command and control systems and other sites, conducted in a joint operation by Air, Navy and Land forces. On 9 April 2003, the statue of Saddam Hussein at Firdos square in Baghdad was toppled. After 21 days of major combat operations, the old regime was no longer able to exercise control over the territory of Iraq (Dale 2008, pp. 18–24). In all three cases, the objective to destroy, disarm, disrupt and render the adversary impotent had been achieved at the operational and tactical levels. There was simply no need for social sciences inquiries, or necessity to win hearts and minds of the population. However, along with the changing character of the conflict, traditional approach could not be sustained any longer. Increasing ineffectiveness of 'shock-and-awe' in Iraq and Afghanistan forced the US and allied forces to redesign their strategic approach and make the cultural turn in following counterinsurgency operations.

The human terrain system

To crush and annihilate the enemy with modern weapon systems was not sufficient to achieve political objectives in Iraq and Afghanistan. The socio-cultural awareness of indigenous populations required extended knowledge and management of a large quantity of data. The new counterinsurgency doctrine expected every military leader to become a social worker, a civil engineer, a schoolteacher. It was hard to imagine a sheer success from military personnel without appropriate language skills and with only a superficial understanding of Afghan and Iraqi cultures (Eikenberry 2013). At first, Cultural Smart Cards (CSC) were introduced to fill in existing gaps. These wallet-sized notes were aimed to provide a portable orientation to Islam (Federation of American Scientists 2006). However, information and knowledge in CSC oversimplified and overgeneralised sociocultural complexities and produced confusion, fostered ethnocentrism and increased misunderstandings between the people. Thus, the US army assembled the Human Terrain System (HTS) programme, which was aimed at supporting the ongoing counterinsurgency missions by establishing teams consisting of military and civilian personnel with an appropriate educational background.

The fundamental element of the Human Terrain System were teams usually consisting of five to nine members including: Team Leader, Social Scientist, Research Manager and Human Terrain Analyst. The primary objectives of their field research were the following:

1 Provide description and analyses of civil considerations for each district and area of operations.
2 Maintain an understanding of local leadership, how they interact with each other and determine their interests and concerns.
3 Provide specialised assistance to brigade combat teams and battalion projects to facilitate completion, efficiency and social impact.
4 Provide guidance to soldiers regarding how to collect human terrain information to improve their intelligence preparation of the battlefield and reporting efforts.
5 Respond to requests for information from the brigade combat team and battalion (Griffin 2010; Finney 2008).

One of the most crucial ethical constraints was related to the use of knowledge that might inflict the harm to the subjects of study. In 2014 the US Senate Intelligence Committee concluded that since September 2001, the CIA had been using various torture techniques to get information from the suspects. According to the 500-page report, some captives were deprived of sleep for up to 180 hours, exposed to simulated drowning and sexually abused (Hosenball 2014). Before this report came to the public eye, in 2004 a journalistic investigation found out that the US military used *The Arab Mind* as a guidebook for (mis)understanding the Arabs. Written by anthropologist Raphael Patai in 1976, it offensively described Arabs as lazy, sex-obsessed and apt to turn violent over the slightest thing. Many scholars described the book as old and discredited, an example of bad, biased social science (Abukhattala 2004). Nonetheless, it appealed to the military due to its simplicity and superficially coherent perception of the Arab enemy (Whitaker 2004). Therefore, the underlying uncertainty among social scientists was that any kind of research produced through HTS-like projects might be handled improperly by intelligence components to inflict harm. However, Montgomery McFate and Kamari Clarke distinguish between providing general knowledge to the army and specific knowledge of persons and activities related to military targeting and intelligence information gathering (McFate 2011, pp. 63–82; Clarke 2010, pp. 208–209). Moreover, the participation of social scientists in the HTS-like projects might have the opposite effect by exposing immoral actions undertaken by GMI agencies and discredit enhanced interrogation as a totally ineffective way of collecting any kind of data and knowledge.

Another related ethical dilemma was the issue of clandestine research. Anthropologist Gerald D. Berreman noted that 'to do research in secret, or to report it in secret, is to invite suspicion, and legitimately so because secrecy is the hallmark of intrigue, not scholarship' (Berreman 1971, p. 396). The major problem with secret research is that it limits the dissemination of knowledge among social sciences and, more broadly, to the society. Scholars have no access to the information and there is no critical peer-review. For instance, Price mentions the PRISP programme, designated to train intelligence operatives and analysts, consisting mostly of university students, who have to keep their affiliation hidden from others on campus (Price 2011, pp. 33–57). It is the GMI that decides what to consider clandestine, how long it should be keep away from the public eye, how this research will be used and by whom. Assuming that anything secret might inflict harm or is unethical blurs the picture behind the reasons why something has been kept secret. Secrecy is a policy employed in pursuit of certain strategic aims or goals, or on behalf of some set of underlying strategic intentions. In due course of field research, the scholar himself/herself might decide not to include some part of information in the research to protect the identity of the subject (Lucas 2009, pp. 168–174). Another type of research might require secrecy as a methodological prerequisite. For instance, psychologists conducting blind or double-blind experiments prevents subjects from compromising the validity of research outcomes. Although clandestine research might be an unpleasant product associated with GMI-funded projects, scholars engaging themselves in this type of work should be fully aware of potential constraints and limitations.

The war in Iraq and Afghanistan generated intense debates about the role of social scientists at the battlefield and regarding cooperation with GMI. Even though social sciences and social scientists do not provide a panacea for all the problems related to warfare, it is cooperation rather than opposition, that might ameliorate the suffering, atrocities and brutality that the victims experience during the conflicts. Furthermore, it is against the notion of social sciences to deny access to the knowledge generated by social scientists. Hence, the concluding part of this chapter presents and discusses three possible ways of cooperation between social scientists and GMI, namely: at the academic and think tank level, the strategic and political level and at the operational and tactical level.

The three models of cooperation

To facilitate interaction with various societies it is necessary to acquire sufficient knowledge. The first model is based on a comprehensive educational process and research. Teaching is what most social scientists actually do, whether is it teaching in public or private universities, or in military academies. Improving the cognitive skills of soldiers, their social and cultural understanding, is a fundamental and primary way to decrease the chances of mistakes and conflicts with the local population. The purpose is to educate soldiers about the diversity of other cultures, and existing differences and commonalities. Among various skills, one should be able to distinguish basic concepts, norms and values in the surrounding environment. The value of knowledge about history and language can surpass the hard power approach. Lack of communication can lead to a lack of cooperation and difficulties in peaceful and productive coexistence. In addition, knowledge of the 'self' might be another possible solution. Cultural literacy is about understanding our own individual cultural patterns and cultural norms. It can promote cultural relativism and reduce biased and ethnocentric perspectives of other societies. There is, however, a significant difference between cultural knowledge and cultural awareness. To become culturally aware, one needs to immerse him or herself into the culture of the host society. Scholars in academia will not teach how to feel another's pain and grievances, feel how threatened another might feel, or to understand how our own actions appear to others. Cultural understanding and cultural training are often either too academic, with no provision for concrete application, or are purely factual, providing only a 'shopping list' of cultural facts or stereotyped instructions, just as in the case of CSC (Odoi 2005). The divergence of theory and practice regarding the cultural understanding of culturally different people should be the focus of further research with the aim of reducing the existing gap. Finally, it is important to remember the potential limits of the profession. A soldier is a soldier, not anthropologist, sociologist or psychologist. He or she receives a specific type of training and develops a specific way of thinking and personality that might constrain perception and understanding of culturally distant societies (Dixon 1994).

The Camelot affair demonstrated that it is difficult to achieve competent scientific transdisciplinary cooperation. Gabriel Almond, one of the participants in a public debate about the Project, explained his view by stating that he is not arguing that scholars who act as consultants to government agencies or who receive contracts for research from government agencies thereby lose their independence of judgment. On the other hand, I cannot go along with those who feel that this kind of pattern of participation has no effect whatever on the independence of the social scientists (Robin 2001).

Government-supported projects usually aim for a specific type of research in support of specific political or military objectives, with a concrete vision for the end result. Some projects might be conducted under confidentiality, without possibility to publish the outcomes. On the other hand, having an influence on and support from governmental institutions might shape the decision-making process that can mitigate negative consequences of political or military actions, and ultimately produce more benefit than harm. The change is possible only from inside-out, rather than outside-in. Thus, the second model focuses on political and military decision makers.

Advising the leadership can be a challenging and sometimes daunting task. However, they make decisions about the use of force and it is their decisions that have ramifications on the lives of people. The past gives a plethora of examples when politicians lack sufficient sociocultural knowledge. A former secretary of defense Robert McNamara once noted: 'I had never visited Indochina, nor did I understand or appreciate its history, language, culture, or values. When it came to Vietnam, we found ourselves setting policy for a region that was terra incognita' (McFate 2005, pp. 42–48). Similar perception of other cultures had been seen in Iraq. In May 2003 Paul L. Bremer, former head of the Coalition Provisional Authority (CPA), issued a CPA order no. 1 banning all but the lowest members of Baath Party from holding any governmental positions. The promulgation of CPA order no. 2 disbanded the Iraqi army leaving over 250,000 Iraqi men out of a job, on the streets, angry and armed. Bremer later explained that his intent was to show to Iraqi people that the Saddam regime is gone and will never return. His order might not have directly supported the insurgents, but he certainly inflamed and accelerated its rise (Kaplan 2014, pp. 74–75). These are just two glaring examples of politicians disregarding the complexities of the security environment in war-torn societies.

The case of Herbert Goldhamer and his involvement in Korean War negotiations also shows how difficult it is for a social scientist to be an effective advisor for policymakers. Despite Goldhamer's presence at the initial stage of talks, American negotiators established their own vision of a simplified, mono-dimensional enemy. On the other hand, the case of Geoffrey Gorer and Ruth Benedict, along with their suggestions to keep the Japanese emperor intact, implies the possibility of having a positive influence on the mind of political elites. There are also two other related risks of advising decision-making groups. First, members of a core group rarely make any attempt to obtain information from experts who can undermine the course of action that had already been established, and ignore or neglect an alternative course of action suggested by the outsiders. Furthermore, as a member of a core group, there is a social pressure on the social scientist to suppress deviational points of view and to remain silent if their own beliefs cannot match up with those of the rest of the core group (Janis 1972, pp. 10–43). Second, there is a question of time pressure. One former special advisor elaborated that 'most cabinet ministers are pretty busy (…) You will look for those opportunities where you can grab hold of them for five minutes (…) your ability to sit down and talk for any length of time is somewhat limited' (Hazell 2014, p. 13). Group thinking and time pressure are part and parcel of political and military institutions. They often impede independent critical thinking, which is likely to result in irrational and dehumanising actions directed against the out-groups. In general, aforementioned models are basically about making trade-offs between academic and governmental work. Participation and cooperation with these institutions should be decided by each individual using his or her own sense of ethical responsibilities. Any academic code of conduct should not prohibit from such engagement. Otherwise, it shifts from an ethical guide to mono-dimensional moralisation struggling with accurate and true, or less accurate and true rules and principles.

The last model of cooperation with the GMI is the most controversial. The Human Terrain System provoked many ethical dilemmas: clandestine research, unintended support of kinetic and targeted killings and the use of knowledge in enhanced interrogation techniques. But at the same time, it is impossible to avoid an amalgamation of social scientists with politics and military conflicts. To ignore this fact is to remain in the position of the three wise monkeys: see no evil, hear no evil and speak no evil. It is to hide behind the walls of academia, to limit the access of social sciences inquiry by the GMI, to accept the brutality of war and political terror as routine, normal, even expected. On the contrary, the neutrality of scholars should obligate them to identify, denounce and oppose all the wrong-doings inflicted during and after the war. Hence, it is a question of when and with whom to engage and cooperate, rather than to cooperate at all. Although the Human Terrain System became the main locus of the debate, there is much more beyond it. Social scientists could contribute their knowledge and expertise in the post-conflict, nation-building and peacekeeping operations. For instance, memory of atrocities is creating a sense of 'unfinished business' that again justifies the resort to violence. In order to mitigate the grievances, to rebuild the trust and will of cooperation between the people, social scientists could work with non-governmental institutions or participate in international committees to resolve such issues. What is more, the internal organisation of international peacekeeping or peacebuilding missions is challenging by itself. It is a complex multitasked, multilateral, multidimensional and multinational/multicultural environment. It requires a different type of mindset and different skills from traditional military attitudes (Rubinstein 2008). Thus, such missions create a unique opportunity for a social scientist to find mechanisms and tools facilitating mutual understanding, respect and successful cooperation and coexistence with the host society.

Concluding remarks

Social scientists as human beings, citizens and all other roles they perform, are engaged in ways that go beyond disciplines and their principles. Various codes of ethics in the academic world are trying to establish 'ideal' rules for scholars, in their own way they are utopian examples of pure science separated from the corrupted outside world. There is no doubt that such guiding principles are desirable and necessary. However, at the same time, they cannot turn into struggling moralisation that impedes realisation and the purpose of social scientists, namely, to serve the society. All three mentioned models of cooperation have some flaws, they are not perfect, but there is no perfect solution in a conflict environment. Sometimes there is no one right way of saying what is seen because there may be no one right way of seeing it. The three models proposed a common ground on which it would be possible to agree or work on, and to focus on mitigating the suffering of the people, rather than debate whether it is ethical to help them. However, further research is necessary to improve the ways and mechanisms that could allow unconstrained cooperation between the world of academia and governmental, military and intelligence agencies.

References

Abukhattala, I. 2004, 'The New Bogeyman under the Bed: Image Formation of Islam in the Western School Curriculum and Media', In J. Kincheloe and S. R. Steinberg (eds), *The Miseducation of the West: How Schools and the Media Distort Our Understanding of the Islamic World*. Westport, CT: Praeger, 153–170.

Andrade, D and Willbanks, J. H. 2006, 'CORDS/Phoenix. Counterinsurgency Lessons from Vietnam for the Future', *Military Review*, 86(2): 77–91.

Benedict, R. 1946, *The Chrysanthemum and the Sword: Patterns of Japanese Culture*. New York, NY: Houghton Mifflin.

Berreman, G. D. 1971, 'Ethics, Responsibility, and the Funding of Asian Research', *Journal of Asia Studies*, 30(2): 390–399.

Chung, Y.-W. 2004, 'Leaflets, and the Nature of the Korean War as Psychological Warfare', *The Review of Korean Studies*, 17(3): 91–116.

Clarke, K. M. 2010, 'Toward A Critical Engaged Anthropology: Diversity and Dilemmas', *Current Anthropology*, 51(Supplement 2): S301–S313.

Conetta, C. 2002, *Strange Victory: A critical appraisal of Operation Enduring Freedom and the Afghanistan War*, Project on Defense Alternatives, Research Monograph no. 6. Cambridge: Commonwealth Institute, 30 January.

Cravens, H. 2012, 'Column Right, March! Nationalism, Scientific Positivism, and the Conservative Turn of the American Social Sciences in the Cold War Era', In M. Solovey and H. Cravens (eds), *Cold War Social Science. Knowledge Production, Liberal Democracy, and Human Nature*. New York, NY: Palgrave Macmillan, 117–137.

Dale, C. 2008 'Operation Iraqi Freedom: Strategies, Approaches, Results, and Issues for Congress', *CRS Report for Congress*, no. RL34387, 28 March.

Dixon, N. 1994, *On the Psychology of Military Incompetence*. London: PIMLICO.

Dye, T., Zeigler, H., Schubert, L. 2011, *The Irony of Democracy: An Uncommon Introduction to American Politics*, 15th Edition. Boston, MA: Wadsworth Cengage Learning.

Eikenberry, K. W. 2013, 'The Limits of Counterinsurgency Doctrine in Afghanistan: The Other Side of the COIN', *Foreign Affairs*, 92(5): 59–74.

Federation of American Scientists. 2006, *Iraqi Cultural Smart Card*, viewed 3 April 2015, http://fas.org/irp/doddir/usmc/iraqsmart-0506.pdf.

Finney, N. 2008, *Human Terrain Team Handbook*. Fort Leavenworth, KS: Human Terrain System.

Gorer, G. 1943, 'Themes in Japanese Culture', *Transactions of the New York Academy of Sciences*, 5(5), Series II: 106–124.

Griffin, M. B. 2010, 'An Anthropologist among the Soldiers. Notes from the Field', In J. D. Kelly et al. (eds), *Anthropology and Global Counterinsurgency*. Chicago, IL: University of Chicago Press, 215–231.

Hazell, R. 2014, *Being a Special Advisor*. London: University College London, School of Public Policy, The Constitution Unit.

Hinton, P. 2002, 'The "Thailand Controversy" Revisited', *The Australian Journal of Anthropology*, 13(2): 155–177.

Hosenball, M. 2014, 'CIA tortured, mislead, U.S. report finds, drawing calls for action', *Reuters*, 9 December, viewed 4 April 2015, www.reuters.com/article/2014/12/09/us-usa-cia-torture-idUSKBN0JM24I20141209.

Janis, I. L. 1972, *Victims of Groupthink. A Psychological Study of Foreign-Policy Decisions and Fiascoes*. Boston, MA: Houghton Mifflin Company.

Kaplan, F. 2014, *The Insurgents. David Petraeus and the Plot to Change the American Way of War*. New York, NY: Simon & Schuster.

Leites, N. 1951, *The Operational Code of The Politburo*. Santa Monica, CA: Rand Corporation.

Lucas, G. R. Jr. 2009, *Anthropology in Arms. The Ethics of Military Anthropology*. Lanham, MD: Rowman & Littlefield Publishers.

Lynn, J. A. 2003, *Battle: A History of Combat and Culture. From Ancient Greece to Modern America*. Boulder, CO: Westview Press.

McFate, M. 2005, The Military Utility of Understanding Adversary Culture', *Joint Force Quarterly*, 38: 42–48.

McFate, M. 2006, 'Anthropology and Counterinsurgency: The Strange Story of their Curious Relationship', *Military Review*, LXXXVI(2): 1–24.

McFate, M. 2011, 'Reflections on the Human Terrain System During the First 4 Years', *PRISM*, 2(4): 63–82.

Odoi, N. 2005, 'Cultural Diversity in Peace Operations: Training Challenges', *Kofi Annan International Peacekeeping Training Center Papers*, No. 4. https://www.africaportal.org/publications/cultural-diversity-in-peace-operations-training-challenges/.

Price, D. H. 2007, 'Buying a Piece of Anthropology. Part 1: Human Ecology and Unwitting Anthropological Research for the CIA', *Anthropology Today*, 23(2): 8–13.

Price, D. H. 2011, *Weaponizing Anthropology. Social Science in Service of the Militarized State*. Petrolia: CounterPunch and AK Press.

Robin, R. 2001, *The Making of the Cold War Enemy. Culture and Politics in the Military-Intellectual Complex*. Princeton, NJ: Princeton University Press.

Rubinstein, R. A. 2008, *Peacekeeping Under Fire. Culture and Intervention*. London: Paradigm Publishers.

Senate Select Committee on Intelligence. 1977, *Project MKULTRA, the CIA's Program of Research in Behavioral Modification*, Joint Hearing before the Senate Committee on Intelligence and Subcommittee on Health and Scientific Research of the Committee on Human Resources United States Senate, Ninety-Fifth Congress, 3 August.

Solovey, M. 2012, 'Cold War Social Science: Specter, Reality, or Useful Concept?' In M. Solovey and H. Cravens (eds), *Cold War Social Science. Knowledge Production, Liberal Democracy, and Human Nature*. New York, NY: Palgrave Macmillan, 1–25.

Ulleman, H. et al. 1996, *Shock and Awe. Achieving Rapid Dominance*. Washington, DC: National Defense University, Institute for National Strategic Studies.

Whitaker, B. 2004, 'Its best use is as a doorstep', *The Guardian*, 24 May, viewed 6 April 2015, www.theguardian.com/world/2004/may/24/worlddispatch.usa.

10 Ethical dimension of post-heroic and autonomous modern armed conflicts

Błażej Sajduk

Introduction

The present chapter tackles social and ethical consequences of the phenomenon referred to as the dehumanisation of the way in which the so-called 'Western' countries wage wars. This phenomenon consists largely of two strongly interrelated processes. The first one was described in the mid-1990s by Edward Luttwak (1995, pp. 109–122; 1999, pp. 127–139) as post-heroic warfare or de-heroisation of the battlefield. The second aspect is the growing autonomy of the military equipment employed. Both are indirectly the result of a trend shaping the direction of the development of weapons which aims to reduce risks associated with the conduct of military operations by increasing the distance from which the enemy is attacked.

Today, thanks to the development of new technologies, one can observe the next stage of this process. In contrast to earlier phases, now we are witnessing a change of a qualitative nature. The increase in distance of the fighting parties from the war theatre has grown so much that currently one speaks of 'disconnecting' soldiers from the battlefield (Coker 2013, p. 125). This 'disconnection' from the realities of the battlefield is also apparent due to the fact that the West is becoming more and more distant – emotionally and culturally – from the enemies it is fighting (Coker 2007, p. 10). The West finds itself in the process of not only moving away the risk from those conducting military operations, but also removing risk from such operations, which makes it increasingly more difficult to convince the public about the value of courage and sacrifice of one's life in war (Sparrow 2015, p. 389).

The idea of courage is one of the pillars that constitute the idea of a warrior, and currently some personnel of the armed forces have ceased to observe the several-thousand-years-old ideal of the warrior. Christopher Coker aptly pointed out that although ancient warriors and medieval knights fought for their country, they did so as vassals to their monarch, and today 'sacrifice is the key to the warrior ethos. Through it, the bond that the warrior forges with his community, his unit and country, becomes a sacred one' (Coker 2007, p. 5). In the 21st century, so far only for a minor part of the military personnel, a great part of the risk associated with their military activities has been eliminated. A transformation of the warrior, who used to be a person whose constitutive feature was risking their own life for the community, is taking place right now. For the last 20 years a new group of 'warriors' within the armed forces has begun to emerge. Though still defending their own society, they do not, however, have to bear all the negative consequences on the battlefield (like, for example, risk of injury or death) comparable to what their predecessors from earlier periods of history had to deal with. That group includes, among others, those piloting unmanned combat aerial systems (UCAS) or carrying out activities in cyberspace. From this perspective, the idea of bestowing armed machines with autonomy is both the consequence and culmination of the process of moving away from the opponent, leading to the ultimate dehumanisation of the battlefield. Of course, in this context, the fundamental objections are raised by the possibility of transferring decisions regarding human to life to Lethal Autonomous Weapons Systems (LAWS).[1]

This chapter is structured as follows: Section 1 – 'De-heroisation' deals with the consequences of the constantly growing distance between warriors and their enemies. Section 2 – 'Autonomy' tackles the problems of ethical and legal consequences of growing autonomy of lethal weapons. Section 3 – 'Conclusion' contains remarks on the future impact and development of the above-mentioned phenomena.

De-heroisation

It is worth remembering that an ideal of the warrior has been developed within European culture and is an important part of its heritage. This legacy has exerted a profound influence on the social idea of the role that soldiers have played in societies. Certainly, the modern perception of the warrior ideal has been fundamentally influenced by the legacy of ancient Greece (largely Athens) and the medieval culture of chivalry. To understand this role, one should begin by recalling the notion of ethos, which is inseparable from the idea of warrior. Ethos is a 'way of life of a given community, the overall (...) orientation of a given culture, the hierarchy of values that it adopted, formulated explicitly or perceptible in human behaviour' and, what is very important 'ethos is a term that applies to groups and not to individuals'. (Ossowska 2000, p. 8). When attempting to find the roots of chivalric ethos, one should turn to *The Iliad*, in which Homer pointed to a set of features which characterised the elite of Greek warriors during the war. Moreover, the list of these features later became a collection of virtues for medieval knights (Zakrzewski 2004, pp. 23–41). Such attributes as physical strength, valuing one's own honour and desire for fame played a key role, as well as 'courage, which was an indispensable virtue, and cowardice was the greatest insult' (Ossowska 2000, p. 25). The particular social and cultural strength of the chivalric ethos was, among others, grounded in the fact that representatives of this social group recognised the desirability of imitating a specific set of behaviour. Therefore, they, as it was believed, would be able to acquire particular desirable traits such as (above all) respect. In addition, a similar mechanism of imitation functioned regarding members of lower social groups who, having no prospects for knighthood, wanted to pursue the admired virtues. It should be clearly emphasised that the idealised image of a warrior depicted him as different from the rest of the community –

> The warrior is much more than a brave man, and he is much more than a man who loves his profession, though he will be born with what we call a vocation; he will answer the call. To be a warrior is much more than a philosophy of life, it is often something akin to a religious calling, and it involves killing, and killing others with skill (Coker 2014, p. 16)

However, with the development of societies, formation of the nation states, and the emergence of new types of weapons, many of the virtues attributed to warriors and knights underwent profound transformation. The 17th century – the era of religious wars – was a period in which the ethos and principles of chivalry were completely set aside, and the place of medieval and Renaissance knights was taken over by an increasingly unified army. The rapid development of ranged weapons, mainly firearms, made horsemen and armoured knights obsolete. Finally, they lost their importance for contemporary warfare. The primary duty of the soldiers in the uniform state armies (Münkler 2004, pp.74–81) became not individual heroism, but obedience and discipline. Gradually, two co-existing models of heroic attitude started to be observed in the military (Coker 2001, p. 34). The first one, of a general nature, meant the willingness to sacrifice one's life and health for a general and abstract idea, e.g. the nation. The latter form of heroism was the willingness to sacrifice one's life for other people, e.g. fellow soldiers, brothers in arms. It seems that both attitudes are currently in crisis in our postmodern society.

Chivalric ethos, bravery and heroism at the turn of the 20th and 21st centuries are different from the ancient and medieval ideals, which provided the opportunity to choose the moment of death, e.g. in an individual clash with the opponent. Three events seem to be emblematic in the transformation of these concepts. One is the French Revolution, which ushered in the appearance of large national armies on the battlefield – then, ideology became a force able to motivate people to fight, thus weakening personal desire to gain glory on the battlefield. Moreover, the increasing professionalisation of warfare meant that the armed forces began to resemble a rational and anonymous bureaucratic machine, described by Max Weber. Other important milestones were the Industrial Revolution and then World War I, during which battles were fought by the 'soldier masses'. It would be a bitter irony to say that during military operations undertaken between the years 1914–1918, soldiers gained the opportunity to be sacrificed like never before. It is from World War I that we know the tradition of making mass graves for anonymous soldiers – the so-called Graves of the Unknown Soldier. There is no doubt that WWI weakened the individual dimension of heroism, making the ideal distant like never before. As Christopher Coker wryly remarked, that war was a mockery of heroic stories which people learned about at schools (Coker 2001, p. 30).

The remnants of the heroic rivalry could be seen only in the air, where pilots, 'the last knights', were fighting face to face.

The last process which seals the change of the warrior ethos is the shape of societies in contemporary, postmodern and highly developed countries, which Ulrich Beck called *risk societies* (Beck 2002). The inhabitants of the Western world are characterised by an overwhelming aversion to risk, which also has consequences for the way in which wars are fought, because 'if the risk age shapes the way we see the world, it has also begun to reshape the military themselves in terms of their own ethos and training' (Coker 2009, p. 27). In warfare, the results of the above process include limiting the risk of casualties (among one's own soldiers and civilians) as much as possible. It should also be emphasised that treating human life as the highest value during times of war was not a widespread approach in recent history, as exemplified by huge losses on the battlefronts of World War II, not only among soldiers but also among civilians. The 21st century has created and supported two categories of asymmetric warriors: on the one hand, we can see non-state actors, such as suicide bombers (and weapons such as improvised explosive devices); on the other hand, we have military personnel fighting in cyberspace and cubicle warriors.

Today, it seems highly tempting to perceive soldiers as 'public servants', providing services in the form of external security safeguards (private military companies may be seen as the most extreme example of outsourcing external and internal security), which also weakens the traditional understanding of the ethos of the warrior. Postmodern and post-materialist societies want to lead 'sterile' wars and try to avoid physical damage, something that was the foundation of wars for last two millennia (Coker 2008, p. 3). Christopher Coker expressed the essence of the transformation of modern armed forces in a witty wordplay – in his opinion, the values of ancient Greeks are disappearing in a modern army for the sake of the values of geeks (Coker 2013, p. 13). The stakes, however, are higher than it may seem. By '(post-) modernising' war and soldiers and removing risk, fear and heroism, purely human qualities, Western societies dehumanise the battlefield, which may result in the instrumentalisation of war, which will cease to have a human, and therefore, an ethical dimension. Until recently, belligerents shared the same universe, they were connected with the same fate, and now one of the parties is beginning to leave the world of their opponent, and by consequence, their fates are also beginning to look different (Coker 2013, p. 102). Furthermore, 'Robots are taking us even further away from the responsibilities we owe our fellow human beings' (Coker 2008, p. 152). Writing in the context of the crews of the ships which fire self-manoeuvring missiles, Herfried Münkler (2004, p. 164) noted that this is a situation when 'the unity of dying and killing, embodied in the traditional type of soldier, is torn apart. (...) Such war has lost all the qualities of a traditional duel, and, cynically speaking, it is similar to certain forms of pest control'.

Therefore, it seems legitimate to draw an analogy between these members of contemporary armed forces (separated from any risk of physical injury or death) and functionaries employed by the state whose profession is to execute capital punishment (executioners or hangmen – professions well known since ancient times). The latter are not exposed to any risk either, and what is more, they have the legal right to take people's lives. Similarly, cubicle warriors, due to the nature of the environment in which they work, have ceased to embody the traditional set of characteristics of the warrior (among others, courage and bravery), and at the same time meet the criteria to be assigned to another professional group – not soldiers, but bluntly speaking, executioners or hangmen.

Autonomy

To begin with, it is necessary to clarify two closely related terms – automation and autonomy. Automation is the performance of pre-programmed operations by a machine in a specific, unchanging environment (e.g. welding manipulators in car factories) (Wagner 2014, pp. 9–13). A characteristic feature of automation is high predictability of the system, which may be adversely affected by the occurrence of faults. Autonomy is a concept broader than automation, which includes the concept of automation and which may be subject to gradation. Autonomy in the context of machines may mean independence of the system in the selection of appropriate measures to implement the expected results (Arkin 2009, pp. 8–13). In other words, a machine undertakes action on the basis of its own software in a non-predefined number of situations, without direct human intervention.

Taking into account the relation of a man with a machine and the possibility of interference of a live operator in the work of a system, after Human Rights Watch and Harvard Law School's International Human Rights Clinic (2013, p. 3), one may differentiate three weapon systems from the point of view of the level of their autonomy (Scharre 2016, pp. 9–10). First, *Human-in-the-Loop Weapons*, where the system sets the target and carries out the attack only after man has given it orders to do so. Up to a certain point, the system executes actions on its own, then it stops and waits for the commands of man. Second, *Human-on-the-Loop Weapons*, where the system can set targets and attack them and man has the ability to block its actions. The role of the human operator is reduced to monitoring the operation of the machine. And finally, *Human-out-of-the-Loop Weapons*, where the system is capable of selecting targets and attacking them without human intervention.

In the ethical sense, autonomy may refer to deliberate actions of a free subject which can be assigned ethical consequences and responsibility (see Kant 2013). In the case of human activity, responsibility and autonomy are two sides of the same coin. From the ethical point of view, a subject that is defined as autonomous makes decisions independently, sets its own goals and ways to achieve them, thus, it is directly responsible for its actions. Assigning responsibility to anyone else is tantamount to the fact that the subject is not autonomous. Thus, mentally sane subjects may undertake activities of moral character and be held responsible for them.

Autonomisation of arms which are used to fight (which, in fact, fight by themselves) may be a logical consequence of a larger trend in the development of mankind, described by Alvin Toffler (1997), in which humanity has gone through three waves of development determined by the development of technology. Man was thus replaced by animals (agrarian revolution), then machinery (industrial revolution) and now a large part of the so-called Western world is in the process of replacing people with algorithms (digital revolution). The consequence of the process of removing physical human presence is also the removing of the features and proper dispositions of human beings, including the possibility of assigning moral character to human activities, and thus, also the responsibility for one's own actions. Importantly, at the present stage of technological development, solely the responsibility of a cause-effect, and not moral character, can be attributed to machines.

Thus, removing man from the decision-making process has consequences of ethical character, as only from a man one could require ethical perfection and efforts to meet (even the highest) ethical standards, while in regard to (autonomous) machines, it is assumed by definition that this type of armament may be unreliable. At this point one may see two conflicting ways of thinking: the first one, which may be called 'utilitarian', assumes that if a machine will prove to be a little more reliable than man, the issue of ethics of its actions should not be treated as a priority and should be replaced by considerations about its effectiveness. The second perspective puts the ethical dimension into the focus, considering, above all, the question of responsibility for human death. In the case of this way of thinking, it is much easier to identify unambiguous opponents of the development of this technology.

Ronald Arkin represents a moderate utilitarian stance, arguing that it is desirable to create such autonomous robotic systems which, in comparison to humans, would break international humanitarian law to a lesser extent (Arkin 2010, pp. 4–10). According to this author, ethical advantages of LAWS are the reduction of losses among the civilian population and troops, strengthened combat capabilities (*force multiplication*), extension of the battlefield, increased capabilities of soldiers and the ability to quicken their reaction and increase the precision of action. Moreover, banning work on this technology now, before results of scientific research are known, is premature and exposes humanity to lowered standards of protecting the life of non-combatants (Arkin 2013, p. 1). If the LAWS technology is developed so as to meet the standards of international humanitarian law, its fate may be similar to that which befell precise ammunition, which due to the reduction of collateral damage came to be considered a weapon ethically more valuable than standard ammunition (Arkin 2014, p. 36). What is more, robots, not having to protect their own life, could be used in a way that assumes their destruction. They can, for example, first take the attack, and only then decide on an appropriate response along the principle *primum non nocere*. Before taking action, robots will also be able to assess risks to health and life of non-combatants. Their sensors will also be more perfect than human senses, thanks to which robots will be able to see through the 'fog of war'. Finally, robots may be designed in a way that will make them immune to the influence of emotions and allow them to analyse the amount of information beyond human cognitive abilities.

A paradoxical consequence of this reasoning is the logical observation that LAWS, if they are unreliable to a greater degree than man, should be prohibited in accordance with international humanitarian law. If they were to prove an effective solution, they would become mandatory ones (Herbach 2012, pp. 3–20). According to the already quoted Ronald Arkin, the possibility of replacing people by LAWS is real, especially in the context of well-defined situations in which one can apply the principles of *bounded morality* (see Wallach 2005). The domain of operation of these machines would not be intranational conflicts and counter-insurgency operations, but conflicts between the armed forces of two countries which are well marked and easy to distinguish (for instance, if the Democratic People's Republic of Korea attacked the Republic of Korea, the need for the use of atomic weapons could prove inevitable, and it would be more ethical to use LAWS in this case) or very specific tasks, e.g. cleaning of buildings.

The ethical perspective assumes that people have the right not to be killed by an (autonomous) machine (see TechDebate 2013). A minimum condition that must be met in the course of war is that it should be possible to assign clear responsibility for human death. Otherwise, war may turn into, to quote Robert Sparrow (2007, p. 9), pest control. Taking of human life is an extreme situation. In the world of Western culture, according to the Christian tradition and modern achievements of human rights, human life is assigned special dignity. In other words, it is important not only why someone died but also how it happened (Asaro 2014, p. 51). Thus, responsibility for the actions (including, above all, those actions that result in people's death) of those who fight in a war has not only a legal but also an ethical dimension. It is impossible to reduce human responsibility only to the dimension of cause and effect, which is possible only in the case of machines. When it comes to the currently used combat systems, to allow a machine to make a decision to take a human life would mean that the responsibility for such an act would only have a cause and effect dimension. Such an act would not have an ethical dimension. For this reason, the use of LAWS could be considered a sign of a lack of respect for human life. This issue was addressed in the study of the Human Rights Watch under the telling title, *Shaking the Foundations*. The authors pointed out that the consequence of the stance assigning every human life the right to protection is the fact that enabling machines to make decisions about life and death deprives people of their dignity, thus contributing to the devaluation of human life. The solution to the problem of 'accountability gap', according to the authors of the above-mentioned report, is to maintain *meaningful human control*, and only such control can guarantee the protection of the dignity of life (and death) of both civilians and soldiers. In addition, only the maintenance of significant human control enables compliance with international humanitarian law (Human Rights Watch 2013, pp. 2–7).

The consequences of a complete removal from war the ethical elements characterising human action – with humans conceived as subjects capable of moral action – and placing LAWS in their stead may be illustrated by an example cited by Robert Sparrow (2007, p. 16), in which an autonomous machine would intentionally attack surrendering combatants. A robot would use its algorithm to calculate all the possibilities and decide to attack. In this case, assigning responsibility is not a simple matter. If in the place of LAWS there were human beings, the case would be clear – they would simply be accused of a war crime. If the responsibility for the death of non-combatants had only a causal dimension and not a moral one, one could seek an analogy with the currently used weapons systems, comparing them to the existing solutions, e.g. 'fire-and-forget' missiles. In this situation, the question of responsibility does not seem to be overly complicated and can be relatively easily assigned to a man who uses the weapon system.

Assigning moral responsibility to the machine itself – which is hypothetically possible – would entail paradoxical and significant consequences. This would require the creation of a 'machine' (quotation marks intended) capable of experiencing higher states of consciousness characteristic of people, which at the present stage of technological development seems to be a thread present mainly in science fiction. However, assuming the possibility of creating such a system, based on *strong artificial intelligence (AI)*, one should at the same time thoroughly review the concept of reward and punishment, which is an integral part of responsibility. Both categories of reward and punishment assume that the subject may experience the states of joy and suffering. In other words, creating machines able to suffer[2] and thus subject to punishment would mean that the purpose for which they were created (reduction of human death) must be questioned, because sending machines that can suffer to war would be ethically reprehensible in the same way as in the case of humans.

Critics of the use of autonomous weapons raise concerns that the algorithms which run these weapons will never be able to process an infinite number of contexts of the given situation to make the right decision. In other words, the speed of compiling quantitative data can never be transposed onto decisions using knowledge of a qualitative nature (International Committee of the Red Cross 2014, p. 7). A machine will never reach a human level of consciousness. In exceptional cases, the advantage of man above the algorithm may be seen in the fact that human beings are able to appeal not only to their previous experiences, but also the broader context of the situation. In addition, the autonomy of a combat system, by definition, involves the ability of adaptation and learning of the machine, which means that its behaviour will not be fully predictable, and thus, a clear assignment of responsibility will be difficult, if ever possible. Thus, the so-called *responsibility gap* will be created – i.e. a certain 'distance' between the original version of the software and that which arises from the collection of new information by a learning system, which can increase so much that it will be completely impossible to assign responsibility, e.g. to software developers of the original version of the system. The paradox of the situation lies also in the fact that the more autonomous systems become, the less predictable their behaviour could be, and it would harder it to assign any responsibility for their actions (e.g. to a commander who decided to use them in a battle) (Matthias 2004, pp. 175–183).

Conclusion

The process of removing the human factor from all spheres of life, such as manufacturing, transportation, and, above all, armed conflicts, seems to be unstoppable. The trends characterised above will likely become more pronounced and so will the (debatable) ethical aspects of the issue. Most certainly, dehumanisation will not apply to every war waged by the Western countries. However, the trend described above has been on the increase for a long time. The armed forces of Western countries are systematically increasing the distance between themselves and the adversary, simultaneously increasing the precision with which they may attack their enemies. This factor makes modern warfare more humanitarian because of the reduction of the so-called collateral damages.

For 20 years, robotics and its consequences – autonomy of future combat systems – have contributed to the change in perceptions of traditional attributes, formerly considered crucial for soldier's ethos and the very concept of warrior. An initiative of the Pentagon is a good case in point to be mentioned here: in 2013 and 2014 an idea for a new type of award – a Distinguished Warfare Medal – was announced. The medal was to be awarded to UCAS operators and cyber warriors. The rationale for giving the awards, however, was neither a physical presence on the battlefield, nor bravery or courage (US Department of Defense 2013b). The greatest controversy was caused by the fact that, in the hierarchy of awards, the medal was to be worn on the uniform, above the Bronze Star with Valor, which is awarded for showing courage in the face of the enemy, heroism or exemplary service on the battlefield, and above Purple Heart, awarded posthumously or for injuries sustained during military service. Eventually, the Pentagon under pressure from veterans, withdrew from the controversial idea, but in mid-2013 changes were made in the manuals for granting medals which removed the concept of bravery on the battlefield.[3]

A lot of media coverage suggested that UCAV operators may suffer from post-traumatic stress disorder (PTSD). Grégoire Chamayou (2015, pp. 103–105) harshly pointed out that in such a case, operators of unmanned systems do not expose their own lives physically, nevertheless they are exposed to mental losses. Therefore, in postmodern war risks to postmodern warriors had to be transformed as well – from the physical domain into the realm of the human psyche. On the other hand, it is worth recalling the opinion of Peter Lee who believes that moral courage needed to take the life of people one observes 24/7 is needed by UCAV operators more than it was ever needed by pilots of manned aircrafts (Lee 2012, pp. 14, 17). This is due to the fact that the number of sensors that UCAVs have got make its operators much less 'disconnected' from the realities of the battlefield than pilots of manned aircrafts operating at high speeds high above the theatre of war.

The appearance of the UCAV was a part of the process of withdrawing the armies of the highly developed countries from the universe shared with the opponent. A logical consequence and at the same time a revolutionary change crowning this process would be the incorporation of the fully autonomous weapon systems in warfare. From the perspective of criminal and civil law, using such systems on the battlefield will be a challenge, especially urgent in the context of assigning responsibility. In regard to international humanitarian law, the problem will lie in ensuring respect for the principles of distinguishing between combatants and civilians as well as proportionality. The emergence of LAWS in military service will be the last stage in the centuries-old process of moving dangers away from one's own forces, and will be tantamount to dehumanising wars and soldiers, which will at the same time contribute to their 'de-ethicisation'. Together with the people, constitutive elements of human nature, including, among others, responsibility, fear and bravery, will also be removed from warfare. This may contribute to the reification of war, which will cease to have its human and, thus, ethical dimension, and will begin to be considered solely in terms of effectiveness and procedural efficiency. This could also mean the end of the world of values in which we have lived since the dawn of history, and the emergence of a completely new order.

Notes

1   Also referred to as Lethal Autonomous Robots (LARs) or simply as Killer Robots (Heyns 2013, pp. 7–8; see also Human Rights Watch 2012, 2013, 2015).
2   This seems impossible and no longer reflects the strong common-sense understanding of punishment - machines obviously do not suffer and thus they cannot be punished for the acts that they have 'committed'. This view is criticised by Gert-Jan Lokhorst and Jeroen van den Hoven (2012, pp. 145–55).
3   However, a definition of courage/bravery remained (US Department of Defense 2013a, pp. 78, 82). In mid-2014 the US Department of Defense initiated work on a new policy of awards in armed forces, the aim of which is establishing a new approach taking into account the realities of the battlefield in the 21st century, including honouring pilots of unmanned platforms (Harper 2014).

References

Arkin, R.C. 2009, 'Governing Lethal Behavior: Embedding Ethics in a Hybrid Deliberative/Reactive Robot Architecture', viewed 10 July 2016, www.cc.gatech.edu/ai/robot-lab/online-publications/formalizationv35.pdf.

Arkin, R.C. 2010, 'The Case for Ethical Autonomy in Unmanned Systems', viewed 10 July 2016, www.cc.gatech.edu/ai/robot-lab/online-publications/Arkin_ethical_autonomous_systems_final.pdf.

Arkin, R.C. 2013, *'Lethal Autonomous Systems and the Plight of the Non-combatant'*, *AISB Quarterly*, 137(07), viewed 10 July 2016, www.unog.ch/80256EDD006B8954 /%28httpAssets%29/54B1B7A616EA1D10C1257CCC00478A59/%24file/Article_ Arkin_LAWS.pdf.

Arkin, R.C. 2014, *Speaker's Summary. Ethical Restraint of Lethal Autonomous Robotic Systems: Requirements, Research, and Implications*, in: *Autonomous Weapon Systems Technical, Military, Legal and Humanitarian Aspects*, viewed 10 July 2016, www.icrc.org/en/download/file/1707/4221-002-autonomous-weapons-systems-full-report.pdf.

Asaro, P. 2014, *Ethical Issues Raised by Autonomous Weapon Systems*, in: *Autonomous Weapon Systems Technical, Military, Legal and Humanitarian Aspects*, viewed 10 July 2016, www.icrc.org/en/download/file/1707/4221-002-autonomous-weapons-systems-full-report.pdf.

Beck, U. 2002, *Społeczeństwo ryzyka. W drodze do innej nowoczesności [Risk Society: Towards a New Modernity]*. Warszawa: Wydawnictwo Scholar.

Chamayou, G. 2015, *A Theory of the Drone*. New York–London: The New Press.

Coker, C. 2001, *Humane Warfare*. London–New York: Routledge.

Coker, C. 2007, *The Warrior Ethos. Military Culture and the War on Terror*. London–New York: Routledge.

Coker, C. 2008, *Ethics and War in the 21st Century*. London–New York: Routledge.

Coker, C. 2009, *War in an age of risk*. Cambridge: Polity Press.

Coker, C. 2013, *Warrior Geeks. How 21st-Century Technology is Changing the Way We Fight and Think About War*. New York: Columbia University Press.

Coker, C. 2014, *Man at War. What Fiction Tells Us about Conflict, from the Illiadto Catch-22*. New York: Oxford University Press.

Harper, J. 2014 'Drone Pilots, Cyber Warriors might get Medals after all', viewed 10 July 2016, www.stripes.com/news/us/pentagon-drone-pilots-cyber-warriors-might-get-medals-after-all-1.273808.

Herbach, J.D. 2012, 'Into the Caves of Steel: Precaution, Cognition and Robotic Weapon Systems Under the International Law of Armed Conflict', *Amsterdam Law Forum*, 4(3), viewed 10 July 2016, http://conflictandsecuritylaw.org/web_ documents/277-1619-1-pb.pdf.

Heyns, C. 2013 'Report of the Special Rapporteur on Extrajudicial, Summary or Arbitrary Executions', 9 April, viewed 10 July 2016, www.legal-tools.org/uploads/ tx_ltpdb/A-HRC-23-47_en.pdf.

Human Rights Watch. 2012, 'Loosing Humanity', 19 November, viewed 10 July 2016, www.hrw.org/report/2012/11/19/losing-humanity/case-against-killer-robots.

Human Rights Watch. 2013, 'Shaking the Foundations', May, viewed 10 July 2016, www.hrw.org/report/2014/05/12/shaking-foundations/human-rights-implications-killer-robots.

Human Rights Watch. 2015, 'Mind the Gap', 9 April, viewed 10 July 2016, www.hrw. org/report/2015/04/09/mind-gap/lack-accountability-killer-robots.

Human Rights Watch, Harvard Law School's International Human Rights Clinic. 2013, 'The Need for New Law to Ban Fully Autonomous Weapons: Memorandum to Convention on Conventional Weapons Delegates', November, viewed 10 July 2016, www.hrw.org/sites/default/files/supporting_resources/11.2013_memo_to_ ccw_delegates_fully_autonomous_weapons.pdf.

International Committee of the Red Cross. 2014, 'Autonomous Weapon Systems. Technical, Military, Legal and Humanitarian Aspects', viewed 10 July 2016, www. icrc.org/en/download/file/1707/4221-002-autonomous-weapons-systems-full-report.pdf.

Kant, I. 2013, *Uzasadnienie metafizyki moralności [Groundwork of the Metaphysics of Morals]*, Kęty: Wydawnictwo Marek Derewiecki.

Lee, P. 2012 'Remoteless Risk and Aircrew Ethos', *Air Power Review*, 15(1), viewed 10 July 2016, www.airpowerstudies.co.uk/sitebuildercontent/sitebuilderfiles/aprvol15no1.pdf.

Lokhorst, G.J., van den Hoven J. 2012, 'Responsibility for Military Robots', In P. Lin, K. Abney, G. A. Bekey (eds), *Robot Ethics The Ethical and Social Implications of Robotics*. Cambridge-London: The MIT Press.

Luttwak, E. 1995 'Toward Post-Heroic Warfare', *Foreign Affairs*, 74(3): 109–122.

Luttwak, E. 1999, 'Post-Heroic Warfare and Its Implications', viewed 10 July 2016, http://www.nids.mod.go.jp/english/event/symposium/pdf/1999/sympo_e1999_5.pdf.

Matthias, A. 2004 'The Responsibility Gap: Ascribing Responsibility for the Actions of Learning Automata', *Ethics and Information Technology*, 6(3): 175–183.

Münkler, H. 2004, *Wojny naszych czasów [Wars of our times]*. Kraków: Wydawnictwo WAM.

Ossowska, M. 2000, *Etos rycerski i jego odmiany [Chivalric ethos and its variations]*. 3rd ed. Warszawa: WydawnictwoNaukowe PWN.

Scharre, P. 2016, 'Autonomous Weapons and Operational Risk, Centre for a New America Security', February, viewed 10 July 2016, https://www.files.ethz.ch/isn/196288/CNAS_Autonomous-weapons-operational-risk.pdf.

Sparrow, R. 2007 'Killer robots', *Journal of Applied Philosophy*, 24(1), March, viewed 10 July 2016, http://profiles.arts.monash.edu.au/wp-content/arts-files/rob-sparrow/KillerrobotsForWeb.pdf.

Sparrow, R. 2015 'Drones, Courage, and Military Culture', In G. Lucas (ed), *Routledge Handbook of Military Ethics*. London-New York: Routledge.

TechDebate 2013, 'Lethal Autonomous ("Killer") Robots', 21 November, viewed 10 July 2016, www.youtube.com/watch?v=nO1oFKc_-4A.

Toffler, A. 1997, *Trzecia fala [The Third Wave]*. 2nd ed, In Poznań: Polski Instytut Wydawniczy.

US Department of Defense. 2013a, 'Manual No 1348.33', viewed 10 July 2016, www.dtic.mil/whs/directives/corres/pdf/134833vol2.pdf.

US Department of Defense. 2013b, 'Memorandum: Distinguished Warfare Medal', 13 February, viewed 10 July 2016, www.defense.gov/news/DistinguishedWarfareMedalMemo.pdf.

Wagner, M. 2014 'The Dehumanization of International Humanitarian Law: Legal, Ethical, and Political Implications of Autonomous Weapon Systems', *Vanderbilt Journal of Transnational Law*, 47, viewed 10 July 2016, www.law.upenn.edu/live/files/4003-20141120---wagner-markus-dehumanizationpdf.

Wallach, W. 2005 'Artificial Morality: Bounded Rationality, Bounded Morality and Emotions', viewed 10 July 2016, www.realtechsupport.org/UB/WBR/texts/Wallach_ArtificialMorality_2005.pdf.

Zakrzewski, L.S. 2004, *Etos rycerski w dawnej i współczesnej wojnie [Chivalric ethos in past and contemporary wars]*. Warszawa: Trio.

11 European military and dual-use technology transfers to Russia

The impact on European and Transatlantic security

Simona R. Soare

Introduction

Technology is a significant driver of military power. Nations are very protective of their military technology and reluctant to share it even with their closest allies. Military technology transfers, which are very sensitive for geopolitical, security and ethical reasons, are closely monitored nationally and internationally. European military and dual-use technology transfers to Russia after the Cold War are equally sensitive in a transatlantic context, for geopolitical, security and ethical reasons. They have repeatedly been challenged for undermining transatlantic security, for reinforcing critical Russian military capabilities (Jermalavicius and Kaas 2014; Jones 2014; van Ham 2015) and for being a major factor in the shifting balance of power between the West and Russia, in Moscow's favour (Belkin et al. 2013). While the Western media has been very critical of European arms transfers to Russia (Jones 2014), they have not received due academic attention. A number of reasons explain the lack of research on European military transfers to Russia after the Cold War: there is not enough information available about them, they are not considered strategically relevant enough (Jones 2014) and European states use them as carrots for a complex partnership with a difficult Russia.

European military transfers and the full implementation of EU-wide arms control regulations are increasingly important in the current strategic context. The consolidation of the Common Security and Defence Policy (CSDP), debates about the EU's strategic autonomy and changes in the European defence market and industrial and technological base[1] may change European motivations for pursuing arms transfers to Russia (European External Action Service 2016; Ayrault and Steinmeier 2016). The signing of the EU-US Acquisition and Cross-Servicing Agreement (European External Action Service 2016, p. 23) may raise additional barriers for European military transfers to Russia (and others). Growing tensions between the West and Russia over the conflict in Eastern Ukraine, Russia's military posture alongside the Eastern flank of NATO and the Russian indiscriminate bombings in Syria have already raised the spectre of additional European sanctions against Russia (Rankin and Asthana 2016; BBC 2017).

In the context of the current challenges, this chapter aims to contribute to the debate on the impact of military or dual-use technology transfers on the shifting balance of power by looking at the case study of European military and dual-use technology transfers to Russia after the Cold War. The central questions are: what impact have European military technology transfers to Russia had on the deterioration of the security environment in and around Europe since the late 2000s? Did these transfers of military technology to Russia ultimately impact the shifting balance of power especially in fringe areas like the Baltic and Black Seas? What were the European and Russian reasons for pursuing military technology transfers? The argument put forth in this chapter is that, cumulatively, the European military and dual-use technology transfers to Moscow had a significant direct and indirect impact on the shifting balance of power between the West and Russia in Moscow's favour, especially in Northern and Eastern Europe and the Middle East. They also helped consolidate the Russian defence industry, enhancing its role as the second largest global arms exporter. Evidence suggests the most significant impact is in areas of command and control, radars, imaging and situational awareness, armour, targeting systems and electronic components (including dual-use components). The impact was fostered by the disconnect between the European states' geo-economic, neoliberal focus in their engagement with Russia, and Moscow's geopolitical focus.

The chapter begins with a brief theoretical background of the research, identifying three theoretical explanations of European military transfers to Russia. The next section analyses comparatively the European and Russian rationales for pursuing military transfers after the Cold War. I look at the type, volume and quality of European transfers to Russia over this period and analyse the multilateral institutional cooperative context around them. The last section focuses on the impact the European military transfers to Russia have had on the balance of power and the security environment in and around Europe. Finally, I conclude by looking at the challenges that lie ahead for the European arms control regime and European military transfers to Russia.

Theoretical background

Arms transfers, a branch of IR, offers a well-researched foundation of the strategic, economic and ethical motivations and consequences of military transfers. Beyond international sales of weapons, arms transfers are more broadly defined as any activity resulting in the trading, donation or assistance of the development of military or dual-use equipment, technology, design, spare parts, training or know-how (Soare 2012; Krause 1992; Stohl and Grillot 2009). Arms transfers are motivated by the maximisation of strategic and economic gains or influence. Arms transfers literature is based on an enduring debate between liberal and realist schools. The liberals argue arms transfer is essentially limited in scope and duration because investments in the military starve the economy (Ikenberry 2001; Keohane 2005), whereas for realists, power or security maximisation, security dilemmas, status and security perceptions drive the international demand for arms transfers (Jervis 1976; Walt 1987; Snyder 1997). While this traditional divide is important, it proves less than helpful in explaining this case study in which the parties hold diverging perceptions about the importance of arms transfers: European states are focused on the neoliberal approach whereas the Russians are driven by a realist one.

The arms transfer literature offers three potential explanations for analysing European military technology transfers to Russia after the Cold War. The first explanation is provided by the vertical flow of military transfers in the pyramidal structure of the global arms market. First-rank producers, i.e. great powers with technological know-how and industrial capacity to drive technological progress, diffuse military technology to more numerous, less advanced second- and third-rank producers who reproduce and adapt weapons systems and use them as means to develop their own industrial base. However, both European states and Russia are first rank global arms producers.

The second theoretical explanation is that states prefer to be in control of their military supply chain and their defence-industrial technological base. Leading arms producers prefer to develop, build and sell advanced military products 'off-the-shelf' to maintain and control their strategic advantage and cutting-edge technology. This creates customer dependency which is both strategically and economically profitable for the producers. Of the three types of arms transfers – material, design and capacity transfer[2] – first-rank arms producers such as the European states and Russia do not normally transfer advanced military technology (i.e. capacity or design transfers) to one another. Russia's case is an atypical one, though, for the Russian industrial-military complex is very flexible and easily adapts to and incorporates foreign military technology.

The last theoretical explanation concerns the timing of advanced arms transfers in the system. Technological progress can foster revolutions in military affairs (RMAs) through the development of new military technology/ platforms and their new strategic or tactical uses. RMAs generally lead to revolutions in warfare (RIWs) through a restructuring of the organisation, conduct and protocols of warfare to introduce and/or make better use of technological progress. As RMAs occur, advanced arms producers invest less and less in their older military technology and become more inclined to transfer it more easily and widely in the international system. RMAs – and globalisation – accelerate the diffusion of military technology, which further shapes the balance of power in the international system (Keegan 2012; Knox and Murray 2001; Gray 2002). As we are approaching a new revolution in military affairs through the military use of automation, cyber, robotics, artificial intelligence, nanotechnology and bioengineering (Burrows 2016), the first-rank arms producers, such as the leading European arms exporters, would diffuse their old (not their advanced, cutting-edge) military technologies to other states, though not necessarily to other first-rank producers, who would presumably have little use for them.

The three theoretical explanations are very useful tools to understand why and how arms transfers occur. But they appear to be insufficient in explaining the case study analysed in this chapter as they do not fully explain the diverging perceptions European governments and Russia hold about the role and importance of arms transfers for international security and the balance of power. Arguably, a different narrative is needed which bridges the gap between the domestic and international rationales for pursuing military transfers and explains the different perceptions about the role and importance of military technology transfers.

As I will demonstrate in the following sections, a neoclassical realist approach to European military transfers to Russia fully explains the disconnect between the two parties' diverging perceptions about the role and impact of arms transfers which resulted in a shifting balance of power and a deterioration of security in and around Europe. It does so by considering not just the structural aspects of power politics but also the domestic considerations that influenced European and Russian policy concerning the military and dual-use technology transfers and how these policies further impacted the balance of power in Europe.

Military technology transfers:
A comparative analysis of European and Russian rationales

European-Russian relations have come full circle: from the tense days of the Cold War to the developing partnership of the late 1990s and 2000s, to contemporary renewed tense and militarised relations after the conflicts in Georgia (2008), Ukraine (2014–ongoing) and Syria (2011–ongoing). The evolution of multilateral institutionalised cooperation mechanisms between Russia and the West has acted as a facilitator of more depth and breadth of European military and dual-use technology to Russia, especially in the 2000s. The suspension of this multilateral cooperation as part of the 2014 Ukraine-related sanctions against Russia politically complicated and slowed down European military transfers to Moscow. But the discrepancy between the European and Russian rationales for engaging in military technology transfers seems enduring. The Europeans had a geo-economic, neoliberal focus guiding their engagement with Russia, whereas Moscow's was a geopolitical focus.

European defence matters ... but trade comes first!

Against the politically permissive background of a closer institutional cooperative framework with Russia especially after the 1997 NATO-Russia Founding Act (NRFA), the establishment of the NATO-Russia Council (NRC), the EU's pursuit of a strategic partnership with Moscow in the early 2000s and the American reset of relations with Russia in the late 2000s, the European approach to relations with Russia was driven by economic necessity rather than a grand strategic design. Economic reasons and the belief that closer economic and industrial interdependence would help consolidate the strategic partnership were usually behind European military technology transfers to Russia (Belkin et al. 2013, p. 588). European governments' optics on military transfers to Russia was they 'could [not] significantly alter regional security dynamics' (Belkin et al. 2013, p. 588), and that 'defense cooperation with Russia provides an important means by which to influence the country's military modernization process, adding that the constructive partnership could benefit from more modern and interoperable Russian military forces' (Belkin et al. 2013, p. 589). These European expectations were never met as the Russian approach continued to be a geopolitical, security competition-driven one in relations with the West.

The enduring cuts in European defence spending after the Cold War (Military Balance, 1991–2016), the negative impact of the economic crisis (Berteau and Ben-Ari 2012) and the peace dividends embraced by the European countries during the last two decades led, in 2014, to a 13.3 per cent cut in European defence R&D in comparison to 2007 levels (Berteau 2015, p. 3; European Defence Agency 2016). Other sources place the reduction in European defence R&D budgets at as much as 29.1 per cent during the same period (Am Cham EU 2017, p. 9). In December 2013 the European Council reiterated 'defence matters' (European Council 2013, p. 1), but defence spending cuts across the continent, and especially in Western Europe, accelerated (Berteau 2015). This restructuring by default of European defence markets eventually threatened the arms market's viability, the integrity of the European supply chain and the global competitiveness of the European arms producers. For the first time since the Industrial Revolution, Europe was faced – and still is – with the prospect of losing its military technological cutting edge within a decade (Drent et al. 2016).

So dire was the situation, the European defence market was faced with a double imperative: to diversify production and to focus on exports to substitute falling defence procurement in Europe. Switching from a state demand-driven production pattern to a market demand-driven production cycle was a difficult change for the European defence market, which increasingly pressured European governments for access to politically sensitive markets, such as Russia. This transformation of the European defence market played a significant facilitator role in European military transfers to Russia, especially during the 2000s. The Russian market played an important domestic procurement substitute for a few European states. Since 1991, the European states were the largest foreign direct investment (FDI) contributors in Russia (European Commission 2014). Russia's 140 million people represented Europe's closest largest developing market, especially after 1999 when the Russian economy started to develop. Opening the Russian market served European energy, security and economic interests. European defence contractors were among the largest investors in Russia. In 2007–2008 the European defence consortium, EADS, bought a 5 per cent share of Russia's VTB, a Russian aerospace contractor, for €1.2 billion. The value of other joint ventures, like Italy's Iveco and Russia's Kamaz, remains undisclosed. Russia was providing highly lucrative contracts for European arms contractors, such as the French Mistral contract worth €1.2 billion or the German Rheinmetall €100 million contract for the Mulino training facility.

The cancellation of these contracts determined a combined net decrease of -7.4 per cent in European arms sales revenues in 2015 in comparison to 2013 (Fleurant et al. 2015, p. 2). In 2015, six of the ten largest arms producers and exporters in the world were European states – France, Germany, the UK, Spain, Italy and the Netherlands – which held a common share of global arms transfers of 23 per cent, less than Russia's 25 per cent market share. European companies are increasingly less dominant in the Top 100 global arms producers. Both Germany and France are exporting significantly less arms nowadays than in 2006–2011 (Fleurant et al. 2016, p. 3). In 2015, with over 25 companies in the top 100 largest arms producers, the Europeans amounted to just 26 per cent of total arms revenues and Russia's 11 defence companies amounted to 10.2 per cent. France and the large trans-European defence consortia were the hardest hit by cancellations of military transfers to Russia after 2014. The German defence industry was the least affected, German arms sales still rising by 9.4 per cent in 2015 in comparison to 2013–2014 (Fleurant et al. 2015, p. 2).

Leading European arms exporters to Russia by number of military technology licenses since 1991 were Italy, Poland, the Czech Republic, France and Germany (SIPRI 2016). By revenue, the most lucrative Russian contracts went to France, Italy and Germany, which were in the position to sell high-end technological licenses for aircrafts (avionics, engines, radar, imaging, targeting; maritime patrol aircraft; attack helicopters), warships, engines and other electronic parts, missile technology, armour, situational awareness, including military intelligence satellites, missiles and command-and-control systems (SIPRI 2016). French companies, Thales and Saffron, sold a number of high-tech licenses to Russian defence contractors in the late 2000s in the fields of avionics, radars and missiles. Thales alone is said to have sold over 1,000 military licenses to Russia since the mid-2000s. Similarly, Italian defence producer Iveco supplied Russia with armour technology and armoured vehicles. In 2007–2008, European governments approved the sale of 1,144 military and dual-use licenses to Russia (Council of the European Union 2008, pg. 203–6) as well as arms sales amounting to €18.5 million. By 2010, they were selling 1,325 licenses to Russia with European arms sales to Moscow worth over €54 million (Council of the European Union 2011, p. 224–227).

However large the European arms transfers to Russia are, they reveal a particular pattern: most transferred military licenses are for parts and electronic components rather than for complete military platforms. The European sanctions against Russia have significantly undermined Russian defence contractors' access to engines and electronic components for aircrafts and radars. YAK-152 trainer aircraft's development was postponed because the German producer of the aircraft's engine was not issued a governmental permission for the license transfer (de Larrinaga 2016). Russia is also facing significant problems with the development of its own actively electronic scanned array (AESA) radar which is intended for deployment on Russia's 4th and 5th generation Sukhoi and MiG multirole fighters. Several critical components of AESA were procured in Europe and their sale is now prohibited by the sanctions regime (Johnson 2016).

Even when Russia is buying military platforms 'off-the-shelf' from Europe, as was the case with the Mistral warships built by France's DNCS or the armoured vehicles developed by Italy's Iveco, they are in relatively few numbers. The Russian government is interested in design and capacity transfers from Europe to reinforce its own military-industrial capacity. Often, Russian authorities negotiate contracts for European design and capacity military transfers for initially small numbers of 'off-the-shelf' military platforms, but with very large production plans as part of a European-Russian joint venture localising production in Russia. The Russian Army only committed to buy 60 Iveco Lynx armoured vehicles; the remaining vehicles were to be built in Russia. Part of the initial 2011 contract was an Iveco-Kamaz joint venture producing 2,500 Iveco Lynx light multirole armoured vehicles in Russia. The production plans stalled because Iveco did not receive all necessary authorisations (Belkin et al. 2013, p. 596).

This particular Russian pattern of military procurement, negotiating the sale of small numbers of military platforms with advanced military technology deployed on them, against the promise of future joint venture producing large numbers of the respective weapons, also raised concerns in Europe if the Russians were going to pursue reverse engineering with European military technology. Intellectual property rights for the military technology sold to Russia and proper control over third-party sales were extremely important aspects of European-Russian defence cooperation. European arms producers, who are interested in maintaining their industrial production capacity and their technological cutting-edge, as well as partially substituting the falling domestic demand, are more interested in material transfers to Russia rather than capacity ones, including production licenses for their military technology.[3]

However effective in cancelling or delaying European military and dual-use technology transfers to Russia (the number of instances where European licenses transfers to Russia are refused by European governments is on the rise after 2014), the sanctions regime continues to be very porous, especially with respect to contracts signed before 2014 and dual-use technology transfers. Despite the sanctions regime, in 2016 the German government allowed a German-Russian joint venture that will produce heavy transport vehicles, including for military use, to be built in Russia and Belarus. This comes at a time when the Russian army is increasingly investing in developing several heavy land robotic systems (Novichkov 2016).

Overall, Russia's defence industry – the second-largest global arms exporter – only amounts to approximately 1 per cent of the European global market share. Russia is not among the five largest clients of any European state (Fleurant et al. 2016). However, European defence contractors have different exposure to the Russian market. French company Thales counts Russia in its top three largest export markets (Rettman 2015). This makes the preservation of defence industrial bases an economic, political and domestic necessity in some of the more exposed states and incentivises them to adopt a neoliberal stance on defence transfers. Against the economic background of the 2000s, even smaller states, like Poland, that are now threatened by Russia's renewed assertiveness and military posture saw the economic and political benefits of military and dual-use technology transfers to Russia. But 1 per cent market share for the European defence industry is rather insignificant. This led many experts to question the impact targeted European sanctions against the Russian defence industry would have (Jones 2014). Equally, it undermined criticism in some European countries that the sanctions were seriously affecting the European defence market.

Russia's great power ambitions and foreign military transfers

Russian interests in military and dual-use technology transfers from Europe were mainly geopolitical and geostrategic. Three rationales can be identified: to reconfirm Russia as a global military great power and offset Western military capability and political influence in areas of strategic interest for Moscow; to refresh the Russian national defence industry's capacity and technological expertise; and to diversify Russia's industrial base to enable it to regenerate its own military capability and maintain its global arms market share.

As mentioned above, Russia is the world's second-largest arms exporter, and dominates some geographical markets like India (over 70 per cent), China (over 59 per cent) or Vietnam (over 93 per cent) (Fleurant et al. 2016). The Asian market – the fastest growing defence market in the world – accounts for roughly 70 per cent of Russian arms exports in 2015–2016 (Connolly and Sendstad 2017, pp. 9–15). Unlike Europe, the Russian defence industry is inherently export-oriented as military exports were an effective source of revenue that ensured its survival during the bleak 1990s and even the early 2000s. The situation only changed with the current Russian state arms armaments programme (2007–2020) worth an estimated €700 billion (Gilbert 2015), but even this is temporary and uncertain under Russia's current economic recession. Despite the political imperative for the Putin regime to achieve the goal of 70 per cent modern military equipment in the Russian armed forces by 2020, recent figures indicate that 'the rate of modernization is actually slowing, with numbers down in aircraft, helicopters, radars, combat vehicles or ship deliveries' (McDermott 2016). The Russian President already warned the Russian defence industry will need to diversify and refocus on exports beyond 2025 in order to maintain its revenues (Zudin and Forrester 2016).

Moscow's preferred military and dual-use technology transfers from Europe were based on a geopolitical approach to global politics and relations with Europe. Since 1991, the evolution of Russian military and dual-use technology transfers from Europe has signalled a Russian selective engagement strategy, which accelerated in the mid-2000s, stimulated by Russian intentions to catch-up with the West technologically and militarily to reaffirm itself as a global great power.

For Moscow, military power and a strong defence industry are considered defining attributes of a great power. Transfers of technology from European producers were driven by a Russian desire to possess the military capability to offset the Western military power and freedom of action in areas of strategic interest for Moscow.

Having a fairly flexible defence-industrial complex that is able to absorb foreign military technology quite easily, Russian arms producers were interested in joint ventures involving design and capacity transfers of European technology through localisation of production in Russia on a majority of its arms purchases from Europe: 'in its arms deals with western countries, Moscow has prioritized acquisition of modern military technology and the licenses to reproduce it' (Belkin et al. 2013, p. 593).

Between 2006 and 2012, France's Thales (e.g. thermo-imaging and targeting systems for Russian T-90 tanks and Damocles surveillance and targeting pod for Russian jet fighters), Sagem (e.g. electro optical infra-red systems for Russian attack helicopters) and Italy's Iveco (e.g. Iveco LMV Lynx armour vehicles) all localised production of advanced military technologies in Russia. The French-developed advanced Damocles surveillance and targeting pod is now used on Russian Sukhoi Su-30-MKMs, on modernised Su-27SM Flanker fighters, Su-24M2 Fencer swing-wing strike aircraft, the Su-35-1 Flanker and the Su-34 Fullback. Sagem's electro optical infra-red systems is deployed on Russian Ka-52 attack helicopters. Sagem's integrated equipment and communications gear is also currently used by the Russian infantry and land forces.

Russia's new Kalibr long-range missile, which is currently deployed with the Russian Black Sea fleet in Sevastopol and playing an active role in the Russian intervention in the Syrian war, has benefited from over a decade of European missile technology transfers. Russia's new Armata tank is benefiting from a significantly improved armour, turret guidance and thermo-imaging targeting system because of military technology transfers from France and Italy.

The Russian interest in buying three Mistral warships from France was mainly driven by their advanced command-and-control technology for amphibious assaults (Rettman 2015) and the training of some 400 Russian sailors on board each warship (Birnbaum 2014). While the Mistral deal did not include NATO-used command-and-control technology, Moscow particularly insisted on acquiring the SENIT-9 combat information system used by the French Armed Forces (Belkin et al. 2013, p. 593) which is compatible with NATO battle command-and-control military technology. In 2015, Russia introduced its first automated command-and-control (C2) – developed on the basis of European military transfers – to its ground forces brigades located in the Western Military District (McDermott 2013).

If in the early 1990s Russia could sell a large volume of weapons from its Soviet-era stockpiles, by the mid-2000s the buyer-driven global arms market – and Russia's main clients – had become more sophisticated. Ballooning defence budgets in China and India, as well as in the Middle East, meant that customers tended to buy more advanced military technology. Transforming its defence sales into more political influence with its large clients, such as China, will increasingly be difficult for Russia if it remains largely cut off from the Western defence-industrial market for a long time. In 2016, India published a tender for a 4th generation multirole fighter which, despite close Russian-Indian military cooperation, was sent to several European arms contractors, including Grippen and Lockheed Martin, as the Indian government was more interested in the newer and more effective Western AESA radars.

Since 2008, Russia launched a reorganisation and modernisation process of its armed forces, including through the regeneration of conventional military capability (IISS 2009–2016). As a result, Russia focused on five priorities in its military and dual use technology transfers from Europe: automation, command-and-control, enhanced accuracy and precision, armour and electronic parts. In 2015, Europeans transferred 351 licenses to Russia – half of those in pre-sanctions years – of which 28 were military technology licenses (France provided 24 of the 28 European licenses), 23 were aircraft-related licenses (including unmanned aerial vehicles, aero-engines and aircraft equipment, related equipment and components, specially designed or modified for military use), 17 were imaging or countermeasure equipment licenses and 11 were non-specified military electronic licenses (Council of the European Union 2016).

The Russian military modernisation process led to an increase in Russian military capability, readiness, mobility and effectiveness. Command-and-control, readiness and mobility and Russian A2/AD capabilities created effective Russian 'A2/AD bubbles' in the Baltic, Black Sea and Eastern Mediterranean (Breedlove 2016; Conely and Rohloff 2016; Bugajski and Doran 2016a, 2016b; Mitchell et al. 2015; Monaghan 2016; Baranowski and Lété 2016). Russia's military modernisation also included a major change in military doctrine with the implementation of a non-linear warfare – also known as hybrid warfare in Western literature – which marked a qualitative and quantitative improvement in Russian military capability and willingness to use military force to achieve political and strategic goals (Lucas and Pomerantsev 2016; Giles 2016; Lanoszka 2016; Renz and Smith 2016).

The experience of Russian observers participating in NATO and European military exercises, Russia's direct military cooperation with NATO and European armed forces between 1997 and 2007 also had a significant impact on the development and use of Russian command-and-control, with significant improvements being visible since the Russian military intervention in Georgia (2008). Between 2006 and 2014, the German government was reported to have sponsored 'several programs aimed at promoting defense reform in Russia, in particular to foster modern defense planning techniques and democratic control of the military' (Belkin et al. 2013, p. 597). In 2015, Russia launched a federal-level, integrated command-and-control system, the National Defense Control Center (NDCC) in Moscow for the modernised and streamlined structure of its armed forces. The German-built state-of-the-art training military facility in Mulino, Russia (the first of several initially planned such Russian training facilities), which was closed in 2015 following the European sanctions against Russia, was one of the most advanced training facilities in the world. It included

> live combat simulation, commander training simulation, marksmanship at modern firing ranges [for a reinforced mechanized infantry brigade]. The centerpiece of the facility [was] a so-called Live, Virtual, and Constructive, or LVC simulation network, which Rheinmetall says, – promises to set a new standard in military training (Belkin et al. 2013, p. 594).

According to the neoclassical realist narrative of military transfers to Moscow, Russian domestic considerations for pursuit of European military transfers were just as important as the strategic and economic ones. Russia's rentier-based defence sector and its political loyalty to the Putin regime were and continue to be a powerful factor in defence budgeting and defence cooperation in Russia. Domestic considerations became increasingly important since 2014 when the Western sanctions were imposed on Russia for its illegal annexation of Crimea and its subsequent involvement in the conflict in Eastern Ukraine.

The Russian defence industry, one of the most technology-intensive sectors of the Russian economy, is one of the leading industries 'through which Russia is integrated with the global economy' (Connolly and Sendstad 2017, p. 25). Since 2014, the Russian economy has been increasingly affected by the sanctions, by falling oil prices and limited access to international capital markets (Connolly and Hanson 2016; de Galbert 2015; Connolly et al. 2015). But it is increasingly rallying around the Kremlin's import substitution policy (i.e. the 2014 Sochi process) which emphasises Russian self-reliance, autonomy and a sustainable, cutting-edge domestic defence supply chain. The import substitution policy proved to be a significant legitimation and consolidation of Russian rentier effects in the defence sector as well as the latter's political loyalty towards the Putin regime after 2014 (Connolly et al. 2015). In 2016, the Russian defence budget was slashed by only 5 per cent whereas other civil sectors (e.g. health, education and social security) saw cuts of up to 15 per cent. The 2017 Russian budget featured a significant drop in defence spending from 23.7 per cent of overall federal budget (approximately 4.7 per cent of GDP) to just over 17.5 per cent (approximately 3.3 per cent of GDP) by comparison to cross-the-board budget cuts of 27 per cent (Hille 2016; Buckley 2016). Financial pressures are increasingly evident in the disagreement between the Russian Finance and Defence Ministries regarding the scale of state funding required to ensure defence modernisation to 2025 as well as the defence modernisation priorities (McDermott 2016).

The security impact of European military and dual-use technology transfers to Russia

The European military and dual-use technology transfers to Russia have had both a direct and indirect impact on the balance of power between the West and Russia and on European security more generally. First, since 2005 European military technology transfers have had a significant impact on enhancing Russian military capability and effectiveness. This impact is most noticeable on Russian command-and-control, radar, imaging, surveillance and targeting, lethal automated systems and armour technology. European military and dual-use technology transfers can be identified along the development chain of many qualitative improvements in Russian military technology over the past decade.

The use of Russian UAVs for intelligence, surveillance and reconnaissance (ISR) in Eastern Ukraine since 2014 to coordinate artillery fire against Ukrainian mechanised infantry units exemplifies improvements in Russian command-and-control systems and in combined arms warfare, a distinct feature of the Western network-centric-warfare. The creation of the centre in Moscow that ensures integrated command-and-control for all Russian armed forces, including for planned and snap military exercises, equally indicates progress in Russian military readiness and coordination. The Russian Armata tank is now able to penetrate the armour system of any old-generation infantry fighting vehicle (IFVs) currently deployed in Europe, without being as vulnerable to European anti-tank inventories. Last but not least, European states would struggle to field enough main battle tanks in case of an aggression on the Eastern flank of NATO, as Germany, the UK and the Netherlands have all significantly reduced their main battle tank numbers in the last decade. Russia, however, would be able to mobilise and deploy up to 22 manoeuvre battalions in their Western Military District alone, 14 of which are mechanised infantry, tank and armour battalions (Shlapak and Johnson 2016, pp. 5–6).

Second, the improvements in Russian military technology that were facilitated by the European military transfers have significantly shifted the balance of power in Northern and Eastern Europe and the Middle East over the past decade. The balance of power along NATO's Eastern flank, from the Arctic, to the Baltic, Black and Mediterranean Seas has been altered in Russia's favour by a combination of Russian non-linear warfare (i.e. hybrid warfare), A2/AD bubbles in Kaliningrad, Crimea and Syria and improved and enhanced Russian military capabilities deployed along the flank. In the

Baltic and Black Seas Russia now enjoys local military superiority (Radin 2017, p. 29; Blank 2017) and 'the overall correlation of forces [is] dramatically in Russia's favor by a margin larger than the acceptable 1-to-3 ratio' (Shlapak and Johnson 2016, pp. 5–6). Additionally, 'Russia (...) maintains superiority in the local military balance across NATO's entire eastern flank' (Hicks et al. 2017, p. 7). Unlike Russia's 22 manoeuvre battalions, European and NATO forces can only deploy 12 manoeuvre battalions, 7 of which are light, not fully mobile, with substantially less armour, artillery and fires and no main battle tanks (Mastriano 2017; Radin 2017; Shlapak and Johnson 2016, pp. 5–6; Conely and Rohloff 2016; Bugajski and Doran 2016a, 2016b; Mitchell et al. 2015; Baranowski and Lété 2016). The balance of power in the Black Sea is such that Russian S-400 air defence systems in Crimea cover two-thirds of the area; Russian Kalibr missiles deployed on submarines in Sevastopol can reach all the way into Syria and anywhere in Eastern Europe. With Russia's A2/AD bubbles in Kaliningrad and Crimea, the West is increasingly losing its air superiority and dominance in Northern and Eastern Europe. This is an unintended but important consequence of the different European and Russian approaches to defence transfers since the end of the Cold War.

Third, the European military transfers to Russia, alongside Russian observers in NATO and European military exercises, Russian participation in joint military exercises and Russian monitoring of Western military engagement in operations worldwide (especially the Balkans in the 1990s and Afghanistan and Iraq in the 2000s) have enhanced Russian awareness of the Western military planning and operations. This deep understanding of Western joint military operations has been instrumental in Russia's new military doctrine of non-linear warfare, which exploits Russia's renewed confidence in its conventional and nuclear capabilities and the West's military vulnerabilities – created by network-centric warfare, the Western technology/information-dependent battle platforms and the globalisation-accelerated diffusion of precision weapons worldwide.

The strategic advantage offered in the 1990s by the American RMA in network-centric warfare based on the system of systems – i.e. linking the weapon platforms with sensors and command-and-control centres via integrated information systems – has eroded during the last decade. The US and its European allies are increasingly regionally challenged from the Baltic Sea to Eastern Europe, the Middle East and South and East Asia (Mastriano 2017; Radin 2017; Conely and Rohloff 2016; Mitchell et al. 2015; Baranowski and Lété 2016). Non-linear or hybrid warfare (Giles 2016; Freier et al. 2016) may well be the current RIW and it is already causing a change in the Western protocols of war. NATO's 2014 Readiness Action Plan (RAP), the deployment of four multinational allied battalions in Poland and the Baltics and a multinational brigade in Romania, the creation of the NATO Counter-Intelligence Centre and the long-term allied deterrence and defence posture adaptation measures, the renewed EU focus on consolidating its CSDP and closer EU-NATO cooperation offer clear indications of the wide Western consensus that the balance of power in Eastern Europe has shifted and Russia is now posing a direct threat to member states on the Eastern flank. Western strategies of countering (Russian) hybrid warfare, the role of cyber and Special Operations Forces (SOFs) and the Western change in the approach of conflict (stages) already signal a significant change in the protocols of war. Further change lays ahead as the US implements its technology-intensive, transformative 3rd offset strategy and the EU develops space-based ISR systems to enhance its situational awareness.

Therefore, European military and dual-use technology transfers to Russia have directly or indirectly helped shift the balance of power between the West and Russia. They have determined significant change in the protocols of war; they helped Russia regain confidence in its conventional military capability, which has been instrumental in Moscow's newfound willingness to use military power to achieve strategically and politically-desirable results, particularly in its near abroad; and they pose ethical and security dilemmas for Europe and the West.

Finally, apart from the security implications of European military transfers to Russia, there is also an ethical dimension involved. European-transferred military technology adapted on some of the most advanced military platforms (the T-90 tanks, UAVs, the attack helicopters Ka-52 or attack bombers Su-34 and Su-35) have been used in Eastern Ukraine and Syria. From a geopolitical point of view such transfers are also questionable, since the most advanced military systems have been deployed assertively on the Eastern flank of NATO. European-transferred avionics and imaging are deployed on Russian fighters that fly with disabled transponders over the Baltic and Black Seas and may violate NATO member states' airspace, increasing the risks of incidents and escalation of conflicts. These are all widely acknowledged to jeopardise European security.

Conclusions

This chapter argues that a neoclassical realist narrative of European military transfers to Russia after the Cold War offers a more comprehensive explanation of European states and Russia's diverging perceptions on the role of arms transfers: European states were driven by a neoliberal approach that favoured trade and treated military transfers to Russia as carrots that would shape Russian military and political transformation after the Cold War. While Europe wanted Russia to become an economic and security partner, the latter continued to perceive the West – i.e. NATO and the EU – as its main national security threats (Ministry of Foreign Affairs of the Russian Federation 2016). Russian authorities regarded such transfers as tools that could be transformed into geostrategic and geopolitical gains. A neoclassical narrative of arms transfers also offers a deeper understanding of the link between the European arms transfers to Russia and the resulting shift in the balance of power in Russia's favour in Northern and Eastern Europe and the Middle East.

The hope and idealism shared by many European decision makers that defence cooperation with Russia would help shape its transformation away from assertiveness, as well as the economic and political need to substitute domestic defence demand with exports, led many European governments to pursue lucrative military transfers to Russia. The conflicts in Georgia, Ukraine and Syria were irrefutable signs Russian assertiveness was not alleviated by defence cooperation with Europe. Great power status and the unchallenged privilege to fully control its sphere of influence in the former Soviet space are strategic goals for which the Kremlin will not hesitate to use military power preventively.

The EU Ukraine- and Crimea-related sanctions against Russia have been renewed on a six-months rolling basis. In December 2017, the European states extended these sanctions, the lifting of which is conditional upon the full implementation of the 2015 Minsk Agreements. As the last three years have shown, the sanctions against the Russian defence industry are not comprehensive: they are limited to three Russian defence contractors and apply to contracts signed after August 2014. European states' reported arms transfers for 2014–2015 continue to show many new military and dual-use technology transfers to Russia, albeit their political approval has become somewhat complicated. Dual-use technology transfers in particular are more difficult to control, despite the Europeans' adherence to the Wassenaar Arrangement.

As the civilian market-driven progress in automation and artificial intelligence gathers pace in the years and decades ahead, a comprehensive modernisation of the European arms control mechanisms and a stricter observance of national reporting and transparency regulations in arms transfers is needed to ensure that sensitive European technology with significant military application will not be transferred to Russia. A consistent EU-wide approach to military transfers to Russia is needed. Cyber, automation, robotics, artificial intelligence, bioengineering and nanotechnology are among the sensitive technologies of the future. The EU and NATO should both play an important monitor role to prevent sensitive military and dual-use technology transfers to Russia.

Notes

1 Examples of upcoming changes include the proposed European Defence Fund, the Preparatory Action on defence-related research, Permanent Structured Cooperation (PESCO) and others.
2 There are three types of arms transfers: (a) material arms transfers, i.e. the transfer/ sale of military platforms or hardware 'off-the-shelf'; (b) design transfers, which are the appropriation of production and design technology along with the military hardware; and (c) capacity transfers, i.e. design transfers accompanied by the appropriation of the know-how, production facilities and/or technologies for the development and production of the respective military capabilities.

3 This trend of European arms market does not apply only to Russia. France's €8 billion contract with India for the sale of 36 Rafale fighters proves European governments are reticent to transfer military design and capability to their non-European customers.

References

Am Cham EU. 2017, 'Security and Defence: Together for European Growth', Brussels: American Chamber of Commerce to the EU, 8 March.

Ayrault, J.M. and Steinmeier, F.W. 2016, 'A Strong Europe in a World of Uncertainties', viewed September 9, 2016, http://statewatch.org/news/2016/jul/de-fr-strong-europe-eu-security-compact.pdf.

Baranowski, M. and Lété, B. 2016, 'NATO in a World of Disorder: Making the Alliance Ready for Warsaw', Washington, DC: Atlantic Council, March.

BBC. 2017, 'Syria war: G7 fails to agree sanctions on Russia after "chemical attack"' *BBC News*, 11 April.

Belkin, P., Mix, D. and Nichol, J. 2013, 'Recent Sales of military equipment and technology by European NATO allies to Russia', *Current Politics and Economics of Russia, Eastern and Central Europe*, 28(5/6): 587–629.

Berteau, D.J. 2015, 'European Defence Trends: Briefing Update', Washington DC: CSIS, 5 January.

Berteau, D.J. and Ben-Ari, G. 2012, 'European Defense Trends 2012: Budgets, Regulatory Frameworks, and the Industrial Base', Washington, DC: Report of the CSIS Defence-Industrial Initiatives Group, December.

Birnbaum, M. 2014 'European countries are selling arms to Russia while condemning it over Ukraine', *The Washington Post*, 17 June.

Blank, S. 2017, 'Are the Baltic States Really Indefensible?' Atlantic Council Commentary, 1 February, viewed 1 May 2017, www.atlanticcouncil.org/blogs/ukrainealert/are-the-baltic-states-really-indefensible.

Breedlove, P. 2016, 'Testimony in front of the House Armed Services Committee', 26 February, viewed 1 May 2017, www.eucom.mil/doc/35166/general-breedlove-house-armed-services-committee-transcript.

Buckley, N. 2016, Russia's renewed might rests on weak economic foundations', *Financial Times*, 31 October.

Bugajski, J. and Doran, P.B. 2016a, 'Black Sea Defended: NATO Responses to Russia's Black Sea Offensive', *Strategic Report no 2*, Washington, DC: CEPA.

Bugajski, J. 2016b, 'Black Sea Rising: Russia's Strategy in Southeast Europe', *Strategic Report no 1*, Washington, DC: CEPA.

Burrows, M.J. 2016, 'Global Risks 2035: The Search for a New Normal', *Atlantic Council Report*, Washington DC: Atlantic Council.

Conely, H.A. and Rohloff, C. 2016, 'The New Ice Curtain: Russia's Strategic Reach to the Arctic', *CSIS Report*, Washington, DC: Centre for Strategic and International Studies, 27 August.

Connolly, R., Galeotti, M., Skyner, L., Chance, C. and Wood, A. 2015, 'Sanctions on Russia: Economic Effects and Political Rationales', London: Chatham House, 30 June.

Connolly, R. and Hanson, P. 2016, 'Import Substitution and Economic Sovereignty in Russia', *Research Paper*, London: Chatham House, June.

Connolly, R. and Sendstad, C. 2017, 'Russia's Role as an Arms Exporter. The Strategic and Economic Importanceof Arms Exports for Russia', *Research Paper*, London: Chatham House, March 20.

Council of the European Union. 2008, 'Notices from European Union Institutions and Bodies. Council Tenth Annual Report according to Provision 8 of the European Union Code of Conduct on Arms Exports', (2008/C 300/01) *Official Journal of the European Union*, 13 December, pp. 203–206.

Council of the European Union. 2011, 'Notices from European Union institutions, bodies, offices and agencies. Council Eleventh annual report according to article 8(2) of Council Common Position 2008/944/CFSP defining common rules governing control of exports of military technology and equipment', *Official Journal of the European Union*, 54(C 9), 13 January, pp. 224–227, 400, 402, 406.

Council of the European Union. 2016, 'Notices from European Union institutions, bodies, offices and agencies. Council Seventeenth annual report according to article 8(2) of Council Common Position 2008/944/CFSP defining common rules governing control of exports of military technology and equipment', *Official Journal of the European Union*, 59(C 163), 4 May, pp. 286–290.

de Galbert, S. 2015, 'A Year of Sanctions against Russia – Now What? A European Assessment of the Outcome and Future of Russian Sanctions', *CSIS Report*, Washington, DC: CSIS, October.

de Larrinaga, N. 2016, 'Russia's Yak-152 trainer completes maiden flight', *IHS Jane's Defence Weekly*, 29 September.

Drent, M., Landman, L. and Zandee, D. 2016, 'A New Strategy – Implications for CSDP', Amsterdam: ClingendaelInstitute, June.

European Commission. 2014, 'European Union, Trade in goods with Russia', Brussels: European Commission', viewed 11 August 2016, http://trade.ec.europa.eu/doclib/docs/2006/september/tradoc_113440.pdf.

European Council. 2013, 'European Council 19/20 December 2013 Conclusion', Brussels: General Secretariat of the Council, EUCO 217/13, 20 December.

European Defence Agency. 2016, 'National Defence Data 2013–2014 and 2015 (est.) of the 27 EDA Member States', Brussels: EDA, June.

European External Action Service. 2016, 'European Union Global Strategy. Shared Vision, Common Action: A Stronger Europe. A Global Strategy for the European Union's Foreign and Security Policy', viewed 5 October 2016, https://eeas.europa.eu/top_stories/pdf/eugs_review_web.pdf.

Fleurant, A., Perlo-Freeman, S., Wezeman, P.D. and Wezeman, S.T. 2016, 'Trends in International Arms Transfers, 2015', Stockholm: SIPRI, February.

Fleurant, A., Perlo-Freeman, S., Wezeman, P.D., Wezeman, S.T. and Kelly, N. 2015, 'The SIPRI Top 100 Arms-Producing and Military Services Companies, 2014', Stockholm: SIPRI, December.

Freier, N., Burnett, Ch.R., Cain Jr, W.J., Compton, C.D., Hankard, S.M., Hume, R.S., Kramlich II, G.R., Lissner, M.J., Magsig, T.A., Mouton, D.E., Muztafago, M.S., Schultze, J.M., Troxell, J.F., and Wille, D.G. 2016, 'Outplayed: Regaining Strategic Initiative in the Gray Zone', Carlisle Barracks, PA: US Army War College Press, viewed 19 September 2016, https://info.publicintelligence.net/USArmy-Outplayed.pdf.

Gilbert, S. 2015, 'Russian Military Modernization', *Draft General Report*, Brussels: NATO Parliamentary Assembly, 24 March.

Giles, K. 2016, 'Russia's "New" Tools for Confronting the West. Continuity and Innovation in Moscow's Exercise of Power', *Chatham House Research Paper*, London: Chatham House, March.

Gray, C.S. 2002, 'Strategy for Chaos. Revolutions in Military Affairs and the Evidence of History', London: Frank Cass.

Hicks, K.H., Conley, H.A., Sawyer Samp, L., Oliker, O., O'Grady, J., Rathke, J., Dalton, M., and Bell, A. 2017, 'Evaluating Future U.S. Army Force Posture in Europe: Phase I Report' *CSIS*, Washington, DC: CSIS, 4 February.

Hille, K. 2016, 'Russia prepares for deep budget cuts that may even hit defence', *Financial Times*, 1 November.

Ikenberry, G.J. 2001, 'After Victory: Institutions, Strategic Restraint, and the Rebuilding of Order after Major Wars, Princeton, NJ: Princeton University Press.

IISS. 2009–2016, 'The Military Balance', *Various Yearbooks*, London: Institute for International Strategic Studies.

Jermalavicius, T. and Kaas, K. 2014, 'France and Germany should stop arms sales to Russia', *EUObserver*, 11 March.

Jervis, R. 1976, 'Perception and Misperception in International Politics', Princeton, NJ: Princeton University Press.

Jones, S. 2014, 'Russia has little to lose from arms embargo', *Financial Times*, 22 July.

Johnson, R.F. 2016, 'Russian industry faces problems fielding AESA radars', *IHS Jane's Defence Weekly*, 3 October.

Keegan, J. 2012, 'A History of Warfare', New York, NY: Vintage Books.

Keohane, R.O. 2005, 'After Hegemony: Cooperation and Discord in the World Political Economy', Princeton, NJ: Princeton University Press.

Knox, M. and Murray, W. 2001, 'The Dynamics of Military Revolution, 1300–2050', Cambridge: Cambridge University Press.

Krause, K. 1992, 'Arms and the State: Patterns of Military Production and Trade', Cambridge: Cambridge University Press.

Lanoszka, A. 2016, 'Russian hybrid warfare and extended deterrence in eastern Europe', *International Affairs*, 92(1): 175–195.

Lucas, E. and Pomerantsev, P. 2016, 'Winning the Information war: Techniques and Counter-strategies to Russian Propaganda in Central and Eastern Europe', *CEPA Report*, Washington DC: CEPA, August.

Mastriano, D.V. 2017, 'Project 1721: A U.S. Army War College Assessment on Russian Strategy in Eastern Europe and Recommendations on How to Leverage Landpower to Maintain the Peace', Carlisle, PA: US Army War College, Strategic Studies Institute.

McDermott, R. 2013, 'Russia's armed forces await automated command and control in 2015', *Eurasia Daily Monitor*, 10: 51, 19 March.

McDermott, R. 2016, 'Assessing the Glacial Progress in Russia's Military Modernization' *Eurasia Daily Monitor*, 13: 179, 8 November.

Ministry of Foreign Affairs of the Russian Federation. 2016, 'Foreign Policy Concept of the Russian Federation', Moscow: Ministry of Foreign Affairs of the Russian Federation, 30 November.

Mitchell, W.A., Lucas, E., Grygiel, J. and Zaborowski, M. 2015, 'Frontline Allies: War and Change in Central Europe', *CEPA Strategic Report*, Washington, DC: CEPA.

Monaghan, A. 2016, 'Russian State Mobilization: Moving the Country on to a War Footing', *Chatham House Research Paper*, London: Chatham House, 20 May.

Novichkov, N. 2016, 'New Russian combat UGV breaks cover, Uran-9 readies for service', *IHS Jane's International Defence Review*, 9 September.

Radin, A. 2017, 'Hybrid Warfare in the Baltics: Threats and Potential Responses', Santa Monica, CA: RAND Corporation.

Rankin, J. and Asthana, A. 2016, 'EU leaders fail to agree on threatening Russia with sanctions over Aleppo', *The Guardian*, 21 October.

Renz, B. and Smith, H. 2016, 'Russia and Hybrid Warfare – Going Beyond the Label', *Papers Aleksanteri no. 1*, Helsinki: Aleksanteri Institute, January.

Rettman, A. 2015, 'French eyes for a Russian tiger', *EUobserver*, 25 August.

Shlapak, D.A. and Johnson, M.W. 2016, 'Reinforcing Deterrence on NATO's Eastern Flank Wargaming the Defense of the Baltics', Washington, DC: RAND Corporation.

SIPRI. 2016, 'TIV of Arms Exports to Russia, 1991–2015', Stockholm: SIPRI, viewed 31 October 2016, www.sipri.org/databases/armstransfers.

Snyder, G.H. 1997, 'Alliance Politics'. Ithaca, NY: Cornell University Press.

Soare, S.R. 2012, 'Transferul de armament', in D. Biro (ed), *Relatiileinternationaleco-ntemporane. Temecentrale in Politicamondiala*, Iasi: Polirom Publishing, 102–116.

Stohl, R. and Grillot, S. 2009, 'The International Arms Trade'. New York, NY: Polity Press.

van Ham, P. 2015, 'The EU, Russia and the Quest for a New European Security Bargain', Amsterdam: Clingendael, November.

Walt, S.M. 1987, 'The Origins of Alliances'. Ithaca, NY: Cornell University Press.

Zudin, A. and Forrester, C. 2016, 'Putin tells Russia's defence industry to diversify', *IHS Jane's Defence Weekly*, 9 September.

12 Dilemmas of security and social justice

The Maoist insurgency in India

Veena Thadani

The national security concerns during the Cold War, with its focus on superpower competition and military power, gave way in the post-Cold War era to a new security context. The 'old wars' of the Cold War era were inter-state conflicts, hence the preoccupation with state security. In contrast, the 'new wars' of the 1980s and 1990s have been intra-state wars, with the causes of conflict seen to reside within, in the nature and failings of the state itself (Kaldor 1999). Intra-state wars comprised more than 95 per cent of all conflicts in the early 21st century (Mack 2005, p. viii).

These intra-state conflicts and the threats they posed led also to a new concept of security – the notion of human security. In contrast to traditional notions of security concerned with the survival and stability of the state and its military capacity, this new notion of security is focused on the well-being of people – human security. It encompasses an expanded range of social and economic factors. In *New Dimensions of Human Security*, the Human Development Report of 1994, human security is defined as 'safety from chronic threats such as hunger, disease, repression', that is economic security, food security, health security, as well as personal, community and political security (UNDP 1994, chapter 2).[1] This suggests that in addition to material sufficiency, human security includes freedom from domination and exploitation, thus embracing a broad range of civil and political, economic and social rights (Thomas 2007, p. 109).[2]

Human security: a concept transformed

In the context of a globalising world – growing interconnectedness in many realms – and the heightened concern about international terrorism, notions of human security were transformed. From the original vision of 'freedom from want' and 'freedom from fear', broad-based poverty reduction and development-oriented goals, human security was re-visioned as a narrower, crisis-focused, new security framework. The effects of poverty, internal conflict and political instability were perceived as threats to regional and global security. In a world of increasingly open borders, the effects of poverty and domestic turmoil may not be regionally contained; issues of global circulation emphasised how conflict in one region may destabilise neighbouring societies and regions with sometimes quite significant implications for distant states through terrorist networks or the displacement/migration of people (Duffield and Waddell 2006, p. 8).

The notion that global poverty and underdevelopment may be dangerous and destabilising internationally led to a shift in focus and priorities from the original version of human security, with its emphasis on the material betterment in the conditions of the poor in the developing world, to an agenda to promote the security concerns of the people in the developed countries of the West. This altered perspective is evident in the report on the Study Group on Europe's Security Capabilities (SGESC), *A Human Security Doctrine for Europe*, in which it is stated: 'the whole point of a human security approach [...] is that Europeans cannot be secure while others in the world live in severe insecurity' (SGESC 2004, p. 10).

The approach to development was also re-oriented, with security concerns fused with development 'towards the rebuilding of toppled states and, in order to stem terrorism, towards addressing popular disaffection in strategically defined areas' (Duffield and Waddell 2006, p. 3). This view echoes that of the Development Assistance Committee (DAC) *Lens on Terrorism* report, with its emphasis on development as a 'strategic tool' in the war on terrorism (Development Assistance Committee 2003). Development is promoted to operate as a security strategy, and the two are inextricably linked: 'no security without development, and no development without security' (United Nations 2005, p. 1). A development agenda, subordinated and re-oriented to security and counter-terrorism concerns – the 'securitisation of development' – has increasingly become a militarised approach to development. It has been described as the '3-D Approach', incorporating diplomacy, defence and development simultaneously (Frerks and Goldewijk 2007, p. 25).

The shift to a security-driven development agenda, with military forces playing a role in development activities – development policy subordinated to security concerns and driven by a militarised logic – reflects a paradigm shift in the notion of human security (Renner 2005). It has been to the 'detriment of the human security approach' and a move to a 'more coercive direction' such that 'the war on terror itself has become a threat to human security by promoting violence instead of eliminating it' (Frerk and Goldewijk 2007, p. 25).

The concern that conditions in third world states – conditions of acute, widespread poverty, social misery, instability and turmoil – may not be contained within the periphery, that in an interconnected and globalising era, they may pose a danger to international security, focused attention on the global implications of internal insurgencies.

The Naxalite/Maoist[3] insurgency in India, which re-emerged in the late 1990s, is presented here as a case study of the risks posed to security – human security, national and international security – by the conditions of widespread acute deprivation and the failure of states to address them, allowing them to fester and serve as incubators of terrorism. The spread of Naxalism is analysed in terms of the social, economic and political structures that produce and perpetuate crushing poverty and exploitation. The Naxalites are waging a violent insurgency in 20 of India's 28 states. The former Indian Prime Minister Manmohan Singh has said that Naxalism is the single most significant internal security threat facing the country.

The Naxalite/Maoist insurgency in India

The Naxalite movement has its origins in the peasant uprising which took place in 1967 in the district of Naxalbari (hence the name Naxalites) in the Darjeeling district of northern West Bengal (Silguri subdivision). The issue sparking the uprising in this area of tea gardens was land – the redistribution of land to landless peasants and tea plantation workers. These peasant and plantation workers in Naxalbari were mainly *adivasi* ('tribal') groups (Santhals, Oraons, Rajbanshis);[4] over 82 per cent were sharecroppers, landless peasants (Louis 2002, p. 51).

The tea plantation workers and other landless peasants were the most deprived segment of rural society; being landless sharecroppers, they were subject to arbitrary eviction. Local land tenure systems and the economic and political power of landowners (doubly privileged in terms of both class and caste) enabled them to impose various kinds of fees and exactions – practices that enriched local landlords and enabled them to accumulate vast fortunes at the expense of their sharecropping tenants and tea plantation workers, who as a consequence were 'brewing with resentment' (Louis 2002, p. 51).

The tea plantation workers and other peasants were mobilised for an armed agrarian struggle – to seize land, overthrow the prevailing 'semi-feudal' conditions in rural society, annihilate class enemies – by a militant group of armed revolutionaries, former members of the Communist Party of India.[5]

The uprising in Naxalbari lasted only a few months; it was suppressed with overwhelming force by the government. Its impact, however, was far reaching; it was the inspiration for the rural poor in many states to launch militant agrarian movements to overthrow the prevailing system and their local oppressors, to seize power and not just land, through guerrilla warfare and violent terrorist attacks on local landlords and the institutions of government (Bannerjee 1984, p. 92; Louis 2002, p. 61).

The 1990s saw the re-emergence of the Maoist movement in the poorest and least developed districts of the states of Andhra Pradesh, Orissa, Bihar, Jharkhand, Chhattisgarh and others – the so-called 'Red Corridor' stretching across central and eastern India. While the geographical reach of the Maoist movement is widespread, estimated to affect 40 per cent of the country, the extent of the insurgency varies by state and district, and is most marked in the areas with a high concentration of India's most deprived and underprivileged – the *dalits* (formerly known as 'untouchables', also known as 'scheduled castes') and the *adivasis* ('tribals'). *Dalits* and *adivasis* comprise about 25 per cent of India's population (over 250 million people) and they are the main base of support for the Maoist movement.[6] The states with relatively high concentrations of *dalits* and *adivasis* in their populations are also the areas of greater Naxalite influence (Report of an Expert Group to Planning Commission 2008, p. 3).

In terms of the actual area of its operation, the Maoist insurgency extends over more territory than the terrorist campaign in Kashmir, the Nagaland insurgency in the Northeast, and the militancy in the Punjab, combined. In terms of the levels of violence, the spread of the conflict and the complexity of the struggle, the Maoist insurgency has been described as the greatest internal security threat to India.

Poverty and the Naxalite insurgency

The Maoists find their recruits among India's poorest and most deprived. The Report of an Expert Group (2008, p. 30), commissioned by the Government of India, asserts a 'direct correlation' between 'extremism and poverty' and concludes that conditions – of acute poverty, deprivation, oppression – create a context for Maoists and other extremist groups.[7] The influence of the Maoists is most evident in districts with the highest rates of poverty, illiteracy and infant mortality (Report of an Expert Group to Planning Commission 2008, p. 9; Borooah 2008, p. 333). In the two states in which Maoist influence is the strongest – Orissa and Chhattisgarh – over 55 per cent of the population live on less than Rs 12 a day (about 40 American cents). The extent of exclusion and marginalisation of the rural peasantry – which makes them susceptible to the Maoist insurgency – is reflected in their conditions of life: in one of the districts of Chhattisgarh, half the people do not know what a train is, do not know what it is like to ride a bus and live without electricity and therefore, without a radio or television (Chakravarti 2008, pp. 42–43).

In addition to the deeply entrenched poverty of the 'scheduled castes and tribes' – who are, in law, entitled to special protections and rights – the *dalits* and *adivasis* are in fact subject to egregious caste-based discrimination and all manner of atrocities in India's age-old, caste-based social order. The above-mentioned Report of the Expert Group refers to the 'life of deprivation, servility, and indignity' of the *dalits* and *adivasis*, the cultural humiliation and the 'structure of oppression' (Report of an Expert Group to Planning Commission 2008, p. 7). The Report documents the growing levels of violence against these marginalised groups by higher caste communities.[8] In the districts where the incidence of atrocities has shown a significant increase between 2001 and 2004, Maoist influence and mobilisation has also shown the greatest increase (Report of an Expert Group to Planning Commission 2008, p. 7). As the Report states: 'Poverty and [...] other factors like the denial of justice, human dignity, cause alienation resulting in the conviction that relief can be had outside the system by breaking the current order asunder' – a recognition of the connection between the conditions of deprivation, grievance and the resort to violence (Report of an Expert Group to Planning Commission 2008, p. 3).

These conditions create an opportunity for Maoists to win the support of the disenfranchised *dalits, adivasis* and the rural poor by addressing their grievances and vulnerabilities – declining wage rates, growing poverty, caste issues – and drawing them into a violent terrorist movement against the state and society that maintains the 'semi-feudal' conditions in which they live. A rural peasant conveys the struggle for dignity and justice that the Maoists proclaim as the cause they champion: 'to save the life of the poor and uphold their dignity' (*garibon ke dharma karam baachal ba, garibon ke izzat baachal ba*) (Louis 2002, p. 259).

In promoting the awareness of injustice and the denial of dignity, the Maoists attempt to change the consciousness of the rural poor and mobilise them to revolt against economic exploitation and social oppression.[9] In his description of how Naxals claim territory and the allegiance of the people, Chakravarti writes,

> [Maoists] show they care, by chasing away forest guards, moneylenders, petty traders, even police, the traditional scourge of tribals and other forest dwellers [...]. When they found they had enough influence, they introduced weapons, started creating local 'dalams' (guerilla squads). Then they created 'sanghams' (groups of active sympathizers) in villages [...] (Chakravarti 2008, pp. 73–74).

The *sanghams*, so-called active sympathisers, are a militia. They form a shadowy army, of indefinite size, providing shelter and logistical support based on the distress – the hunger and anger – which the Maoists draw upon. It is by mobilising these sympathisers that the Maoists have created an infrastructure at the local level, which poses a challenge unlike that of other insurgencies in India.

'State as oppressor'

India's development strategy until the late 1990s conformed to the conception of a 'developmental state' – an activist state engaged in national planning and public investment in the pursuit of economic growth and, rhetorically, socialist goals. State intervention in the economy was justified on the grounds of social objectives, the alleviation of mass poverty.

In 1991, unable to meet the requirements of international creditors, India bowed to IMF/World Bank pressures and adopted the prescribed elements of 'structural adjustment' policies – the opening of India's previously protectionist economy to foreign trade and investment, liberalising imports, privatising state enterprises and substantially reducing government spending and social subsidies in agriculture, health, education, public employment – resulting in profound adjustments in economic and social priorities. These reforms amounted to a major re-direction of the Indian economy, substantially curtailing state intervention, dismantling the state apparatus of development and increasing the role of the market in line with neo-liberal ideas of privileging private enterprise.

The impact of these developments, the restructuring of the state, reflected also the shift in focus from issues of equity and social justice, the broad general welfare aspects of development, to economic concerns of efficiency and economic growth. Issues of distributive justice were eclipsed by the emphasis on economic growth. In a society as highly stratified as India's, in both economic and social terms, economic growth refracted through the prism of such stark inequality has intensified existing disparities. For the large segment of society that remains in conditions of degrading poverty, the neo-liberal project eroded the promise of social justice.

The retreat of the state from the development process led to policies of decentralisation and devolution of power to regional state governments. The poorest areas in which the Maoists are active are rich in mineral resources: 70 per cent of India's unmined iron ore deposits lie in these areas; coal and bauxite are also abundant in these states (Barman 2009).[10] With the neo-liberal reforms of the 1990s and the opening of India's hitherto closed economy, there was an influx of Indian and international companies in pursuit of valuable resources of bauxite, iron ore and coal. The state governments in the region granted mining and land rights to Indian and transnational companies. In Chhattisgarh, the state government signed a 'memoranda of understanding' and other agreements with transnational companies such as De Beers Consolidated Mines, BHP Hilton, Rio Tinto and US companies, such as Caterpillar, which seek to sell equipment to the mining companies operating in the region.

These mining projects have led to the displacement of the rural inhabitants of the area as the powerful corporations move into the tribal area, cut down forests and sell the timber as well as the minerals under the ground (Barman 2009). It is estimated that over 60 million people have been displaced in the period from 1947 to 2004; tribals, who constitute about 8 per cent of the country's population, are 40 per cent of those displaced and *dalits* are 20 per cent (Report of an Expert Group to Planning Commission 2008, p.15). The original claims of the Maoists that they would protect the tribal rural folk from the (relatively) petty manipulation by traders and money lenders has recently changed. The Maoists now claim that they protect the tribals from the land-grabs of corporations, in concert with the state: 'If the Maoists don't protect them, nobody else will' (Chakravarti 2008, p. 30). The Maoist presence in the area serves to obstruct corporate exploitation of the minerals, forests, water and land resources of the tribals and *adivasis* (Navlakha 2006).

The 'development-induced-displacement' of the rural populations of these areas for projects, from which they derive little benefit, and the threat to their environment of mining and construction operations creates an opportunity for Maoists to recruit from their ranks – fertile ground, since the displacement and dispossession greatly exacerbates their conditions of deprivation. In the district of Dantewara in Chhattisgarh, for example, iron ore processing plants dot the landscape, mining approximately $80 billion of iron ore. The largest plant is in Bailadila, in central Dantewara; it is a state-owned enterprise of the National Mineral Development Corporation (NMDC). The five mines in this complex process about 50,000 tons of ore daily. The mines generate heavy pollution, the rivers are contaminated with iron dust and the only sources of fresh water for the use of local villagers and for irrigation are made unusable (Miklian 2012, p. 301).

Projects aimed at 'development', 'growth' and the alleviation of poverty, while generating wealth for some also generate poverty through displacement for the dispossessed (Mathur 2008, p. 2). Padel and Das point out that, 'from tribal people's perspective, all these projects should probably be called *displacement projects* – the word *development* acts as a mask and contradicts their experience' (2010, p. 353).

The Maoists have adapted their strategies in their campaigns to mobilise the poorest and the most marginalised groups in the rural parts of the country under their sway. While the earlier phases of the movement were focused on the exploitation of rural peasants by 'feudal' landlords and money-lenders – as in the 1967 armed uprising of peasants against local landlords in Naxalbari – in the post-1990 phase, there is a shift in strategy to the state as oppressor.

The strategy of the Maoists is now directed against the policies of the government, state and national, and the corporate sector and the plans to establish Special Economic Zones (SEZs) on agricultural land, to introduce commercial crops by multinationals and to the increasing exploitation of forest reserves for mining and industrial expansion. The creation of these SEZs and the granting of mining permits expropriate the property rights of the poor, the common resources which are the basis of their livelihood, amounting to a 'land grab' (Sundar 2011, p. 176).

The Maoist response to these policies has been to mobilise the displaced and dispossessed, the tribal and forest dwellers, against the mining and industrial projects in these mineral-rich areas. The mobilisation is now against 'imperialism' rather than 'anti-feudal struggles against landlordism', in response to the neo-liberal policies of the state in the context of globalisation (Bannerjee 2012, p. 117).

The state strikes back

The Maoists are known for their meticulous preparations, the coherence of their plans – working the lay of the land, the economy and society. Their ideological unity, based on common goals, is adapted to each particular region and situation – in Bengal, extreme rural poverty; in Andhra Pradesh and Bihar, issues of caste and landlord dominance; in Chhattisgarh and Orissa, tribal destitution and displacement; in Karnataka, farmer indebtedness and caste exploitation. New issues are taken up as they arise, such as that of genetically modified seeds, which has become a contentious issue in the cotton growing states of Maharashtra, Andhra Pradesh and Karnataka (Chakravarti 2008, pp. 117–118).

In an attempt to destroy the Maoists web of support, to deny them their assets – food, shelter and sympathisers – the state government launched a counter-insurgency strategy in 2005 in the districts of the state of Chhattisgarh, considered to be the stronghold of the Maoists. The programme, *Salwa Judum*, which, in the local Gondi dialect, translates as 'purification hunt' (but which the government translates as 'peace march') was to co-opt the tribal people of Chhattisgarh into a state-sponsored armed militia to fight against their fellow tribesmen, who may be Maoist sympathisers. The tribal people, potential supporters of the Maoists, were pre-emptively moved to settlements where they could be 'protected' from them (resembling the 'strategic hamlets' programmes in the Vietnam war). The segregation and control of the rural villagers was considered an essential element of a successful counter-insurgency campaign, with the creation of resettlement camps to protect the rural folk from Maoist indoctrination and to destroy any possibility of village-level support for the Maoists (Chakravarti 2008, pp. 17, 55).

To cleanse the countryside of Maoist support, the *Salwa Judum* counter-insurgency strategy set off an internal war, tribal against tribal, a scorched earth policy which included the burning of everything that could not be carried away – food, belongings and such – to deny the Maoists these resources. Trees and bushes were slashed and burned to eliminate the possibility of cover for the Maoists (Navlakha 2006; Chakravarti 2008, pp. 53–58). The tribal villagers, accustomed to living in open habitats, with homesteads separated from each other, have been resettled in camps. Their bare livelihoods, based on farming, tending cattle, gathering forest produce, have been destroyed.

'Operation Greenhunt' is a more recent counter-insurgency operation of the government. Launched in 2009, it calls for the use of police and paramilitary forces in the forests of Chhattisgarh to wage a campaign of targeted assassinations to crush the Maoists (Roy 2011, pp. 45, 132).

These counter-insurgency measures, in addition to the drive to industrialise, the creation of Special Economic Zones, the granting of mining permits result in the expropriation of the property rights of the local people – the common resources that are the basis of their livelihood – uprooting them, causing 'traumatic, psychological and socio-cultural consequences on the affected population [...] in particular the weaker sections of the society [...]' (Indian National Congress 2007).

The counter-insurgency strategy runs counter to the official recognition, in the Report of an Expert Group, that the growth and spread of Maoism is indisputably related to the conditions of acute poverty, deprivation and marginalisation (Report of an Expert Group to Planning Commission 2008). The government's policy fails to address the underlying conditions fuelling the insurgency; it targets the poor, rather than the structural conditions of poverty and socioeconomic exploitation. The Maoists are seen as being similar to the other insurgencies in India – the militancy in Jammu and Kashmir, the separatist violence in the northeast, the (1980s) secessionist movement for

Khalistan (Second Administrative Reforms Commission 2008). The Maoist movement, however, differs from these other insurgencies in that it is not a separatist movement based on ethnic, religious or linguistic identity.[11]

Conclusions: security and social justice

The Indian state, in its neo-liberal incarnation, has adopted policies that focus on economic growth to the neglect of distributive justice. In its policies of land seizures in the mineral-rich areas of Eastern and Central India, in its development-induced-displacement in the Maoist belt, the Indian state has adopted a violent, 'development or perish' approach. In its counter-insurgency policies to crush the Maoist movement, the Indian state itself has become a threat to the security of its citizens. In its failure to address the acute poverty, deprivation and long-standing injustices that fuel the Maoist movement, the Indian state has failed its poorest citizens; it has perpetuated a cycle of violence and counter-violence that undermine both human security and the security of the state. As the Report of the Expert Group stated,

> the Naxalite (Maoist) movement has to be recognized as a political movement with a strong base among the landless and poor peasantry and adivasis. Its emergence and growth need to be contextualized in the social conditions and experience of people who form a part of it ... the fight for social justice, equality, protection, security and local development (2008, pp. 59–60).

Notes

1 The theme of people at the heart of the concept of human security is clear in Kofi Annan's *We the People* (Annan 2000) with its individual chapters on 'freedom from fear' and 'freedom from want'. These themes are reiterated in Annan's *In Larger Freedom* (Annan 2005). *Human Security Now*, based on the work of Amartya Sen on development as freedom and on human capabilities, conceptualises human security in terms of the 'vital core' of human lives – the 'set of elementary rights and freedoms people enjoy' and 'the conditions that menace survival, the continuity of daily life and the dignity of human beings' (Commission on Human Security 2003, p. 4).

2 Human security has been interpreted in different ways with different emphases by various writers. Different proponents of the two core strands of human security focus on different elements. Thus freedom from fear has been seen as an aspect of human security that is humanitarian in its conception – the protection of individuals in situations of violence and is anchored in humanitarian intervention, humanitarian relief, and in tangible achievements such as the land mines treaty in 1997 and the International Criminal Court in 2002 (Thomas 2007, p.109). For a discussion of the disputes over the concept of human security – its ambiguity, conceptual overstretch, the lack of clear differentiation between human development and human security, its limited utility as a policy paradigm, see Martin and Owen (2010).

3 The Naxalites have adopted the 'Maoist' label to identify with the Maoist/Chinese model of revolution with its focus on the rural poor as the main revolutionary

force, rather that the urban proletariat as in the Bolshevik revolution. This armed extremist insurgency continues to be referred to as Naxals/Naxalites/Maoist in public discourse in India; Maoist is the term used in this paper.

4 The term 'tribe' is used in Indian policy discourse to identify groups that are specifically listed in the official Schedule of Tribes, entitled to a system of preferential treatment/affirmative action in education and public employment opportunities. Groups listed as 'tribes' are defined by their geographic isolation, and their social, cultural, linguistic distinctiveness – their 'tribal characteristics', such as 'primitive background', and 'distinctive cultures and traditions'. The *adivasis* (indigenous people) use the bow and arrow as an ethnic symbol, as a mark of 'primitiveness'.

5 In the late 1960s, various factions of the communist movement in India challenged the vision and programme of the dominant party, the Communist Party of India. These factions, advocating the path of agrarian revolution, established the Communist Party of India (Marxist-Leninist) in 1969. Other groups, with differing views on the strategy of revolutionary struggle in the countryside included the Maoist Communist Center (MCC) and the People's War Group (PWG). It was these latter groups, the MCC and the PWG, whose declared objective was to abolish the 'feudal order' in rural India that led the insurgency against the Indian state in the 1990s. They have mobilized the peasantry, particularly the 'tribals' and landless poor in the heavily forested district of Andhra Pradesh, Madhya Pradesh, Orissa, Bihar, Jharkhand, Chhattisgarh, Maharashtra and the conflict has escalated in recent years.

6 *Dalits* comprise about 16 per cent of India's population, about 170 million; *adivasis* about 8 per cent, over 84 million. 80 per cent of *dalits* and 92 per cent of *adivasis* live in the rural areas (Report of an Expert Group to Planning Commission 2008, p. 3).

7 The causes of intra-state conflict in poor and underdeveloped countries in much of the empirical research since the 1990s has focused on theories of 'greed'– that emphasize the economic opportunities in wars – and explanations in terms of 'grievance', that is 'loot seeking' and 'justice seeking' (Collier 2000).

The 'greed' model suggests that states with significant 'lootable' resources – diamonds, other gemstones, minerals, oil – are significantly more likely to experience conflict than countries without primary commodities. In the 'greed' model, civil conflict is seen as driven by the opportunity to acquire power and resources; conflict in poor countries is analysed 'as an instrument of enterprise and violence as a mode of accumulation' (Collier 2000). In this model, Collier and his co-authors explain civil conflict in terms of leaders motivated by the financial gains to be made from conflict situations – a 'self-perpetuating quest for loot'. In this 'greed' view of civil conflict, there is a connection made between conflict and criminality, with civil wars and uprisings described in terms that resemble organised crime, with a focus on rebel finances, extortions and violence. Greed is found far more significant than grievance as an explanation of conflict (Collier 2000).

Attempting to measure motivations such as greed and emotions such as grievance through precise quantitative indicators has obvious limitations; as Collier acknowledges, there are often mixed motives and narratives: 'greed-motivated rebel organizations will embed their behavior in a narrative of grievance' (Collier 2000, p. 1). In their most recent paper, Collier and his co-authors revisit the issue of the causes of civil war. In acknowledgement perhaps of the criticisms of their earlier attempts to measure empirically 'greed' and 'grievance', Collier et al. overturn their earlier results. They present a 'feasibility hypothesis, in which they accord primacy to opportunity – that a rebellion will occur where it is financially and militarily feasible; motivation is now seen as both indeterminate and secondary (Collier et al. 2010).

Explanations of conflict in terms of 'greed' discount the conditions of oppression and injustice, often grounds for grievance, as being the language of protest used by rebel/militant groups to create a sense of victimization – the 'inculcation of grievance' – as the basis for mobilisation and recruitment. Such an approach serves to delegitimise conditions of violent opposition even in the face of genuine conditions of injustice, political repression and the suppression of rights. As Duffield comments, 'the argument and the evidence used to support it ('greed' approach) are a powerful means of delegitimisation and a good excuse for the World Bank and others to pay little attention to critical voices from the South' (Duffield 2001, p. 132).

These conditions of oppression and injustice, dismissed in the 'greed' model, are the explanatory factors in the grievance approach, with the emphasis placed on the underlying structural causes that are the basis of conflict. Poverty, and the factors accompanying it, of social injustice, extreme inequality, inadequate institutions and governance, and often ethnic and religious divisions are seen as creating a context for protest, rebellion, conflict (Stewart 2000). Thus, in the grievance approach, poverty itself is a critical, underlying structural cause of violent conflict (Saether 2001).

8 In *Untouchability in Rural India*, Shah records examples of caste violence, discrimination, and humiliation by higher and intermediate castes against lower castes across rural India. The callous indifference of the authorities provide the Naxals with the opportunity to channel the anger, rage and resentment and bring it to the fore in their mobilisation efforts (Shah 2006).

9 One of the recruits to the Naxal campaign is quoted by Chakravarti: 'Sir, I am from a lower caste. In my village high caste people would not even permit ... they wouldn't even allow us to walk on their shadows. But now I sit on a charpoy above them, and they sit on the ground. Because I am a Maobadi (Maoist). Seeing this, other people have also joined, become part of armed squads and militias' (2008, p. 117).

10 It is estimated that the financial value of the bauxite deposits in one state alone – Orissa – would be about $4 trillion, more than double India's GDP (Roy 2009).

11 The Maoist movement extends over several culturally diverse states, with over six different languages.

References

Annan, K. 2000, *We the People*. Report of the Secretary General, New York: United Nations.

Annan, K. 2005, *In Larger Freedom: Towards Development, Security and Human Rights for All*. Report of the Secretary General, New York: United Nations.

Bannerjee, S. 1984, *India's Simmering Revolution: The Naxalite Uprising*. London: Zed Books.

Bannerjee, S. 2012, 'Reflections of a One-time Maoist Activist', In A. Shah and J. Pettigrew (eds), *Windows Into A Revolution. Ethnographies of Maoism in India and Nepal*. New Delhi: Orient Black Swan.

Barman, A. 2009, 'The Real Solution for Naxalism', *The Economic Times*, 17 November.

Borooah, V. K. 2008, 'Deprivation, Violence, and Conflict', *International Journal of Conflict and Violence*, 2(2): 317–333.

Chakravarti, S. 2008, *Red Sun. Travels In Naxalite Country*. New Delhi: Penguin Viking.

Collier, P. 2000, 'Doing well Out of War', In M. Berdal and D. Malone (eds), *Greed and Grievance: Economic Agendas in Civil Wars*. Boulder: Lynne Rienner.

Collier, P., Hoeffler, A. and Rohner, D. 2010, 'Beyond Greed and Grievance: Feasibility and Civil War', *Oxford Economic Papers*, 61(1): 1–27.

Commission on Human Security. 2003, *Human Security Now*. New York: United Nations.

Development Assistance Committee (DAC). 2003, *Lens on Terrorism Prevention*. Paris: OECD.

Duffield, M. 2001, *Global Governance and the New Wars*. London: Zed Books.

Duffield, M. and Waddell, N. 2006, 'Securing Humans in a Dangerous World', *International Politics*, 43(1):1–23.

Frerks, G. and Goldewijk, B. G. (eds). 2007, *Human Security and International Insecurity*. Wageningen: Wageningen Academic Publishers.

Indian National Congress. 2007, Rehabilitation and Resettlement Bill 2007, viewed 30 April 2017, http://dolr.nic.in/RRBill2007.pdf.

Kaldor, M. 1999, *New and Old Wars: Organized Violence in a Global Era*. Stanford: Stanford University Press.

Louis, P. 2002, *People Power: The Naxalite Movement in Central Bihar*. New Dehi: Wordsmiths.

Mack, A. (ed). 2005, *Human Security Report 2005, War and Peace in the 21st Century*. New York: Oxford University Press.

Martin, M. and Owen, T. 2010, 'The Second Generation of Human Security: Lessons from the UN and EU Experience', *International* Affairs, 86(1): 211–224.

Mathur, H. M. (ed). 2008, *Social Development Report 2008: Development and Displacement*. New Delhi: Council for Social Development.

Miklian, J. 2012, 'The Purification Hunt: The Salwa Judum Counterinsurgency in Chhattisgarh', In A. Shah and J. Pettigrew (eds), *Windows Into a Revolution: Ethnographies of Maoism in India and Nepal*. New Delhi: Orient BlackSwan.

Navlakha, G. 2006, 'Maoists in India', *Economic and Political Weekly (India)*, 3 June.

Padel, F. and Das S. 2010, *Out of this Earth. East India Adivasis and the Aluminium Cartel*. New Delhi: Orient BlackSwan.

Renner, M. 2005, 'Security Redefined', In: *The State of the World 2005: Redefining Global Security*. New York: Worldwatch Institute / Norton.

Report of an Expert Group to Planning Commission. 2008, *Development Challenges in Extremist Affected Areas*. New Delhi: Government of India.

Roy, A. 2009, 'The Heart of India is Under Attack', *The Guardian*, 30 October.

Roy, A. 2011, *Walking with the Comrades*. New Delhi: Penguin.

Saether, G. 2001, 'Inequality, Security and Violence', *European Journal of Development Research*, 13(1):193–212.

Second Administrative Reforms Commission. 2008, Combating Terrorism, 8th Report, New Delhi, viewed 30 April 2017, http://darpg.gov.in/sites/default/files/combating_terrorism8.pdf.

SGESC. 2004, *A Human Security Doctrine for Europe*. Barcelona: Study Group on Europe's Security Capabilities.

Shah, G. 2006, *Untouchability in Rural India*. Thousand Oaks: Sage Publications.

Stewart, F. 2000, *Crisis Prevention: Tackling Horizontal Inequalities*, Queen Elizabeth House Working Paper no 33. Oxford: Queen Elizabeth House.

Sundar, N. 2011, 'The Rule of Law and the Rule of Property: Law-Struggles and the Neo-Liberal State in India', In A. Gupta and K. Sivaramakrishnan (eds), *The State in India after Liberalization*. London: Routledge.

Thomas, C. 2007, 'Globalization and Human Security', In A. McGrew and N. Poku (eds), *Globalization, Development and Human Security*. London: Polity Press.

UNDP. 1994, *Human Development Report 1994. New Dimensions of Human Security*. New York: United Nations.

United Nations. 2005, *In Larger Freedom*. New York: United Nations.

13 Cyber security norms in the Euro-Atlantic region

NATO and the EU as norm entrepreneurs and norm diffusers

Joe Burton

Cyber security has quickly become one of the most discussed and controversial security issues of the 21st century (see Fiddner 2018). Militaries across the globe are investing heavily in offensive and defensive cyber capabilities and many international conflicts now exhibit a cyber dimension, where cyber-attacks are used to control the information environment and degrade the ability of an adversary to communicate. Non-state actors, including the Islamic State in Iraq and Syria (ISIS), are acquiring the ability to conduct damaging cyber intrusions, and the sophistication of malware, computer viruses and tools of cyber espionage are growing rapidly. The strategic, political and legal challenges associated with the malicious use of the internet by states and non-state actors are only just beginning to be addressed and international cooperation on the wide range of issues that the new information age is creating is still in its infancy. It is clear too that there are significant international differences over internet governance, with the US and European visions of a free and open internet standing in contrast to a desire on the part of Russia and China to be able to control information and restrict the use of the internet for activism and dissent.

In the context of these challenges, there has been a growing focus within academic and policy circles on the role of 'norms' in regulating international activity in cyberspace. In the relative absence of cyber security cooperation between the 'great powers', encouraging individual actors within the international system – including nation states and regional organisations – to abide by certain standards of behaviour is seen as one way of mitigating some of the effects of the rapid growth in malicious online activity. As well as the adoption of legal mechanisms to guard against damaging cyber-attacks, the hope is that there will be some sort of collective, social acceptance that certain activities – such as wholesale cyber espionage, or cyber-attacks that target civilians and related infrastructure – should be 'off the table'. Just as in other areas of security – such as the use of chemical, biological and nuclear weapons – norms of behaviour could lead to a mutually constructed environment of restraint and constraint that enhances international peace and security.

This chapter aims to contribute to these debates by examining the emergence of cyber security norms in the European and North Atlantic area and the potential for such norms to be diffused to other world regions. Such a focus on regional norm development is particularly important in the context of the recent failure of the United Nations Government Group of Experts (GGE) process to reach a consensus on its latest report on cyber norms and international law. The chapter proceeds in three parts. The first section seeks to define what cyber security norms are and to categorise them. The second section looks at how norms have been tentatively established in the Euro-Atlantic area through the European Union (EU) and North Atlantic Treaty Organisation (NATO). These organisations, it is argued, have acted as 'norm entrepreneurs' by building support for cyber security standards within and on behalf of their member states. Third, the chapter looks at norm diffusion – the process by which norms are re-established, and to some extent redefined, in different social, cultural and regional contexts. The chapter argues that while cyber security norms are in an early stage of development globally, there has been significant normative progress within the Euro-Atlantic region. Additionally, there are some common global interests that could lead to a 'norm cascade' in other world regions, including the Asia Pacific.

What's in a name? Norms and cyber security in the 21st century

How do we know a norm when we see one? How do we know norms make a difference in politics? Where do norms come from and how do they change? These were the questions posed by Martha Finnemore and Kathryn Sikkink in one of the seminal articles on norms, *International Norm Dynamics and Political Change* (Finnemore and Sikkink 1998, p. 888). Since that article, much progress has been made on answering these basic but important questions. Research on norms and international security has become a major focus within International Relations and statesmen now regularly acknowledge the importance of the normative environment in their day-to-day business; President Obama's reference to the normative dangers of the use of chemical weapons in Syria in 2013 was a prominent example. Normative research and literature has also focused on emerging security challenges and new technologies. The development of the Responsibility to Protect (R2P) norm in the late 1990s, for example, was closely analysed (and influenced) by normative scholars and, in the post-9/11 environment, the need for new norms around counterterrorism, state failure and the use of UAVs (drones) in counterterrorist operations has captured the attention of IR scholars. Now, because of the rapid development of Information and Communications Technology (ICT), and the vast array of political, social and economic effects that have been produced as a result, a body of scholarship is beginning to emerge that pertains to cyber security norms – defined here as expectations of appropriate behaviour that constrain unauthorised access to and/or efforts to disrupt, delay or destroy computers systems.

Finnemore and Sikkink define norms as 'standard(s) of appropriate behaviour for actors with a given identity', and this is a useful starting point for understanding cyber security norms (1998, p. 891). What is appropriate behaviour when using computers and computer networks? The answer to this question will of course differ from individual to individual and from country to country. In this sense, norms are both socially constructed and culturally subjective; what is considered appropriate behaviour online will be different across cultural contexts. Norms are also often slow to emerge, sometimes taking decades to be established, and the result of a complex process of social interaction, advocacy and contestation between actors. Finnemore and Sikkink identify three specific stages in this process, which they call the 'norm life cycle': (a) *norm emergence*, where norm entrepreneurs engage in a persuasion of persuasion, encouraging international actors, and states specifically, to embrace new norms; (b) *norm cascade*, where a critical mass of states agrees to adopt those norms; (c) *norm internalisation*, where norms become taken for granted by states and are no longer a matter of debate (Finnemore and Sikkink 1998, p. 895). A further sub-stage involves *norm diffusion*, where norms are internalised by new states but 'localized and translated to fit the context and need of the norm-takers' (Acharya 2013). This part of the process will be further discussed later in the chapter.

Norms can be categorised in several different ways. There is a distinction, for example, between legal norms that stem from treaties and international law, and political norms, which are voluntary non-binding expectations of behaviour (Rõigas 2015). Another way to categorise norms in the cyber security area is by offense and defence – offensive norms would preclude cyber tools being used to disrupt, delay, destroy or monitor computer networks, while defensive norms would create expectations that actors in cyberspace invest in and are responsible for adequate computer network protection and defence.

Political norms in particular may be well suited to mediating international cyber security behaviour. James Lewis claims, 'Norms remain the best approach because there is too much distrust among competing nations for any legally binding agreement' (Lewis 2014, p. 3). It is expected, following this line of reasoning, that negotiating the terms of legal cyber agreements, monitoring compliance and verifying restrictions on cyber weapons or particularly damaging cyber capabilities/attacks will be more difficult than in conventional areas of security. Even in the area of conventional and nuclear disarmament, treaties are notoriously difficult to enforce. The Non Proliferation Treaty (NPT) of 1970, for example, has been plagued by problems relating to establishing the exact capabilities and intentions of signatories. The spread of dual-use technology, used for both nuclear weapons and energy programmes, has further complicated efforts to reduce proliferation. Given the ease with which states can hide their cyber capabilities, obfuscate any attempt at monitoring and claim that software and hardware development is serving non-malicious objectives, non-codified, more flexible normative frameworks might yield more success than international law, even though they are not legally enforceable.

Even though the internet is still a relatively new technology, there appear to be some common standards emerging that could be adopted by groups of countries (and to some extent already are) that would define and establish what is appropriate and not appropriate when using the internet. Perhaps the most prevalent current approach to norm development in the cyber security area is to apply existing norms of international behaviour to cyber security issues. This is a process by which norms are adapted to fit a new technological imperative. This approach appears to have been the main preference of the US government and is evident in the Obama administration's International Strategy for Cyberspace of 2011 (White House 2011). The strategy underlines that the internet should be free, open, secure, subject to minimal government restriction and that existing norms of international behaviour should apply just as much in the online environment as in the physical world, including freedom of expression and association, respect for property (including intellectual property), respect for privacy, a willingness to take action against criminals and a right to self-defence (including against cyber-attacks) as enshrined in the UN Charter. China too has recognised the principle that existing norms should be applied to cyberspace and that this would be beneficial to China's national interests in certain areas, including the UN Charter norm of non-interference in the affairs of other states, prohibitions and restrictions around the 'use of force' and 'armed attack', and the democratisation of global internet governance (Austin 2016, p. 176).

As well as these general normative concerns, more specific cyber governance norms may be necessary for the ongoing functioning of the global internet in its existing form. The aforementioned US Strategy mentions a number of these 'cyber-specific' norms, including ensuring that the communications system continues to be developed with *Global Interoperability* in mind – that IT systems continue to work across platforms and countries; that countries respect the need for *Network Stability* and *Reliable Access* – i.e. that states do not unnecessarily restrict access to the internet or the flow of information across and between national communications grids; that the principle of *Multi-stakeholder Governance* is upheld, which is to say that governments must include non-state actors in determining how the internet is operated; and that states exercise *Due Diligence in Cyber Security*, meaning that states must have some degree of responsibility when it comes to protecting computer networks from damage and/or misuse (White House 2011, p. 10).

Another way to think about cyber security norms is to consider the motivations for malicious cyber activity and tailor new and existing norms to particular categories of cyber security – most importantly, cybercrime, cyber espionage and cyberwarfare. This categorisation reflects the distinct operational features of each of these fields, the likelihood that cyber security norms will develop a strong sectoral character and that they are likely to evolve at different speeds and over different time-periods.

Cybercrime, which is probably the most prevalent and economically damaging form of online activity, presents several distinct normative challenges that could be usefully isolated and targeted for norm development. Norms are never going to eliminate the financial motivations for online crime, including money laundering, identity theft and bank fraud. However, norms that incentivise international cooperation between law enforcement agencies, and develop cross-jurisdictional mechanisms to hold hackers and cyber criminals accountable for criminal activities, could be useful in providing a greater degree of deterrence against such activity. Cyber norms in the law enforcement area are likely to be distinct from those in the military domain.

Agreements on cyber espionage might also constitute normative frameworks between states. Recent negotiations between the US and China indicate at least some shared willingness to prohibit this kind of activity and the countries made a commitment in September 2015 'that neither country's government will conduct or knowingly support cyber-enabled theft of intellectual property, including trade secrets or other confidential business information, with the intent of providing competitive advantages to companies or commercial sectors' (White House 2015). The same agreement highlighted that both sides were explicitly committed to 'appropriate norms of state behaviour in cyberspace'. This example was emulated in April 2017 by an agreement between Australia and China, which may form the basis for further similar developments in the Asia-Pacific (Prime Minister of Australia 2017).

Perhaps the area where there is the most potential for norms to emerge and, conversely, in which the absence of norms could be the most damaging, is cyber warfare. Here, prohibitions on cyber weapons or a cyber arms control framework might make meaningful contributions to international peace and security. There is no global consensus on what constitutes a cyber weapon, nor indeed on the threshold a cyber-attack would need to reach to constitute an 'armed attack' or 'use of force'. Nevertheless, the growing development of offensive cyber-attack capabilities in militaries across the globe, and the convergence of the cyber domain with more conventional domains of warfare, including air, space, sea and land operations, suggests that cyberwarfare is a growing reality in international politics. It is difficult to tell exactly how powerful a cyber weapon, or sustained cyber-attack, could be. For various reasons, the most capable cyber security actors, the US, Russia and China, have exhibited a degree of restraint when using cyber tools in conflict situations, which suggests that normative constraints may already be a factor in cyber warfare (Valeriano and Maness 2015). However, as the destructive power of cyber capabilities grow, as they are likely to in the absence of any major prohibitions, the pressure to adopt and develop normative constraints is likely to increase. It took some time for norms around the use of nuclear weapons to emerge and various crises fuelled their development, most notably the Cuban missile crisis. If 'cyber weapons' are seen to have even a fraction of the destructive capability of weapons of mass destruction, the development of normative constraints around cyber warfare may accelerate. In particular, mutually acknowledged constraints around the targeting of civilians or related critical infrastructure – such as transport, power, water, energy and health facilities – may constitute norm emergence.

While the emergence of cyber security norms yields some promise, there are various obstacles which would need to be overcome for there to be a 'norm cascade' – the widespread (global) adoption of cyber security norms and widespread (if not universal) consensus on what is and what is not acceptable behaviour in cyberspace. Perhaps the main hurdle, which has been alluded to already, is the desire on the part of some countries to restrict access to the internet, to censor certain websites, and to seek to control the flow of information to their citizens. All countries have some level of censorship, but the level of control in China and Russia in particular, where online media is often manipulated by the state, does not encourage global acceptance of cyber security standards. Norms will likely need to reflect some degree of 'cyber sovereignty' if they are to be globally accepted.

Relatedly, there are ongoing incentives for misuse of the internet in the areas of cyber espionage and cyber warfare. Cyber espionage to obtain proprietary military technology is seen as a way of closing the military gap with America, and cyber-attacks used in warfare are viewed as a mechanism to gain asymmetrical advantages at a tactical, operational and strategic level. These incentives are unlikely to disappear, especially in a period of heightened geopolitical tension. Cyber is still in the process of being established as a field of security and warfare, moreover. The destructive potential of cyber-attacks is still an unknown variable because the peak of technological development in ICT is still over the horizon. States will be reluctant to seek or accept constraining norms that restrict and limit their ability to pursue the full potential of what promises to be the most transformative technology of the 21st century. If history is any guide, non-proliferation regimes develop only after the full destructive potential of new technology is demonstrated and understood through testing in controlled environments. Norms in cyberspace may only be fully acceptable when the limits of the information and communications technology become more apparent.

Another reason norms might be slow to emerge is duplicity – states may agree to abide by certain standards but continue to flout them. The same dynamic has occurred in other areas of security, in the nuclear non-proliferation regime, for example, and there is no reason to suggest that this will not happen in the cyber security sphere. The attribution problem – the ability to attribute cyber-attacks to specific actors – may also constrain norm emergence. If one cannot identify who is responsible for cyber-attacks, if states can cloak their identity, then the incentive to abide by cyber security norms is reduced considerably. There have been considerable advances in digital forensics in recent years that allow for higher degrees of accuracy when attributing cyber-attacks to specific actors. However, normative constraints are dependent on 'back patting' (being praised for good behaviour) and shaming (being publicly castigated for bad behaviour). The level of political and legal deniability that states can maintain around malicious cyber activity arguably reduces the effectiveness of these social dynamics.

Finally, the elevated influence of non-state actors in the cyber security sphere, including terrorist groups, 'hacktivist' groups like Anonymous and businesses and corporations, means that the adoption of norms must be supported by a much wider degree of actors than nation states. ISIS has no incentive to adhere to normative constraints on cyber terrorism or terrorist use of the internet. Similarly, it has proved difficult to persuade the private sector to adopt cyber security standards when there are significant financial or reputational costs involved. Cyber-attacks against Mitsubishi Heavy Industries in Japan in 2011, for example, which led to a loss of data pertaining to the production of advanced military technology, went unreported to the Japanese government for eight months. Norms that mediate the relationship between governments and private actors within their jurisdictions may thus be slower to emerge and more problematic than norms between nation states.

The EU and NATO: 'Norm Entrepreneurs'?

International organisations have been at the forefront of recent attempts to develop security norms and have historically been one of the most effective and influential 'norm entrepreneurs' – actors who seek to change social norms (Sunstein 1996, p. 6). In the area of cyber security, many international organisations are developing a significant normative role. The United Nations, for example, has been involved for over a decade in a process of norm development through the UN Group of Governmental Experts in the Field of Information and Telecommunications in the Context of International Security. Despite the failure of the group to reach consensus in its latest round of reporting, these efforts are likely to continue in some form. The G20 has also taken steps to acknowledge cyber security norms. At its 2015 summit in Turkey, members declared their common commitment 'to the view that all states should abide by norms of responsible state behaviour in the use of [information and communications technologies]', specifically referring to hacking and commercial cyber espionage (Williams 2015). As well as global institutional progress, regional organisations have taken an interest in the development of cyber security norms. The most developed world region in this regard is the Euro-Atlantic area and much of the progress on cyber norms within this region can be attributed to the EU and NATO. This section of the chapter looks at the types of norms established by these two organisations and the complementarity of their work as norm entrepreneurs.

NATO's normative role in cyber security

NATO is not a normative institution per se, but its actions in the field of international security have often set normative precedents that have contributed to norm emergence. In the 1990s, for example, as the organisation became increasingly involved in humanitarian intervention and peacekeeping, most notably in Bosnia (1995) and Kosovo (1999), the alliance influenced the emergence of R2P as a normative framework. The Alliance's military action in Libya in 2011 was justified on the basis of this influential norm and, as the security situation in the country deteriorated in the aftermath of the operation, NATO was criticised for failing to exercise Responsibility while Protecting (RwP), a new normative framework aimed at minimising the adverse effects of such interventions.

In the cyber security sphere, NATO has taken several institutional steps to adapt to the emergence of cyber-attacks as a military and strategic tool in international affairs and the collective actions of NATO's 29 states, which include many of the world's leading military powers, have begun to contribute to the establishment of distinct normative precedents within the North Atlantic region and in the regions in which NATO is operating.

The first stages in this process were largely reactive. They were driven by events, crises and changes in NATO's external security environment that NATO was not actively shaping. In particular, the cyber-attacks against NATO's digital networks in 1999 by Serbian and Chinese hackers, as a response to NATO military action in Kosovo, created concerns within the alliance about non-state actors' malicious use of the internet and placed cyber security firmly on NATO's collective agenda (Denning 1999). These concerns became even more pronounced after 9/11, when NATO leaders acknowledged fears that terrorist groups were intent on developing destructive cyber-attack mechanisms.

Perhaps the most influential event that contributed to NATO's role in cyber security was the cyber-attacks against Estonia in 2007 by Russia-based hackers, which were in response to the removal by the Estonian government of the 'Bronze Soldier', a monument to the Soviet liberation of Estonia. It remains unclear whether the attacks involved the Russian state directly, yet they triggered a sustained debate and series of actions within NATO that relate to normative questions, including: how the alliance should respond collectively to cyber-attacks that reached a certain level of damage, how the alliance could be involved in encouraging states to stop hackers within their jurisdictions using cyber-attacks for political purposes, the threshold at which cyber-attacks should be considered an 'armed attack' or 'use of force', and what responsibilities NATO had under Article 5 (collective defence) to assist one of its members under cyber-attack. In response to these cyber intrusions NATO began to assert itself more actively in shaping the normative environment around the use of cyber-attacks. In 2010, cyber security appeared prominently in the Alliance's strategic concept, an acknowledgment that cyber security was within the purview of the organisation and one of its emerging responsibilities. The Alliance also established the Cooperative Cyber Defence Centre of Excellence (NATO CCD COE) in 2008, a NATO-affiliated research and policy development organisation, and the Emerging Security Challenges division in 2010, focused primarily on cyber security, terrorism and energy security.

This institutional capacity began to lay the foundation for the alliance to make a substantive normative contribution. The NATO strategic concept, for example, referred to NATO's role in enhancing information sharing between NATO members – referring to 'integrating NATO cyber awareness, warning and response with member nations' (NATO 2010). Creating an expectation of behaviour that information pertaining to serious cyber incidents is reported and shared between states is one of the most significant normative concerns in the cyber security sphere. The CCD COE also published The Tallinn Manual on the International Law Applicable to Cyber Warfare in 2013, which is one of the most detailed attempts to look at how existing international law might apply in cyberspace. The manual constitutes an effort to develop legal norms around a range of the most critical cyber security issues, including the launch of cyber operations against critical infrastructure and cyber-attacks targeting enemy command and control systems (Schmitt 2013). This was followed by the release in February 2017 of the Tallinn Manual 2.0, which seeks to complement the original document by examining how international law applies to 'cyber operations' that fall below the threshold of a use of force or armed conflict, including 'state responsibility for operations in cyber space, standards of attribution, the obligation to respect state sovereignty and what the possible responses of victim states to cyber-attacks might be' (Vihul 2017).

The adoption in September 2014 of an enhanced cyber defence policy at the NATO Wales Summit also constituted a normative contribution to the alliance's role in cyber security in the North Atlantic area by stating that cyber-attacks against members could lead to an Article 5 response under the North Atlantic Treaty, meaning that the principle of collective defence would now apply to cyber-attacks. The normative aspect of this policy position should not be underestimated. NATO's policy is that cyber-attacks that reach a certain threshold will be met with a collective response. Individual NATO members would not be on their own in the event of a major cyber-attack. As well as creating a degree of deterrence against such attacks, the principle of collective defence in cyberspace could be emulated elsewhere. NATO's willingness to enact that policy could also set broader normative precedents for cyber warfare – at the present time NATO is not ruling out a conventional response to a cyber-attack, a missile strike on a computer server for example – and the threshold that an attack would need to reach to trigger Article 5 is undetermined.

A further recent normative development within NATO, and one that is closely linked to the collective defence policy, is that NATO has now declared cyberspace to be a domain for military operations, joining the four other warfighting domains of land, air, sea and space. While the focus of this doctrinal development is operational – to enhance planning and coordination for NATO member states and to facilitate the development of new cyber capabilities – it also sets a normative precedent. As Julian Barnes argues, the policy 'will begin a debate over whether NATO should eventually use cyber weapons that can shut down enemy missiles and air defences or destroy adversaries' computer networks' (Barnes 2016). It could also pave the way for more NATO members to develop offensive cyber-attack capabilities to be deployed in conflict situations. As of 2017, NATO's cyber doctrine is defensively orientated and based on protecting its own networks. Finally, in line with the Article 5 policy, the new position sets a potential normative precedent for a cyber-attack being considered an act of war.

Europe and the EU's soft cyber power

In announcing the recent NATO policy shift above, the NATO Secretary General, Jens Stoltenberg, emphasised the importance that NATO work with its long-time partner, the European Union, in developing common approaches to cyber security (Clark 2016). Indeed, the work of these two organisations should be seen as complementary in the area of cyber security, both in terms of their cyber security interests and areas of focus but also in respect to norm emergence and development. While NATO's normative influence is growing in the area of cyber warfare, the European Union has tended to focus on civilian and criminal mechanisms and policies relating to cybercrime and the economic aspects of cyber espionage (as opposed to the military/strategic ones). In this respect, there is a division of labour and responsibilities between the organisations – between NATO's hard power and collective defence approach, based largely on responding to cyber warfare, and the EU's soft power (see Figure 13.1). In fact, norms have been seen as a fundamental role of the European Union in many policy areas, including defence and security, and the EU's normative role is arguably more pronounced, well established and acknowledged than NATO's, particularly in the area of regional integration (Chaban et al. 2015).

The main pillars of the EU's emerging cyber security role are the 2013 EU Cyber Security Strategy and the subsequent Directive on Security of Network and Information Systems (NIS Directive), adopted by the European Parliament on 6 July 2016, which implements some of the main components of the strategy. The strategy emphasises norms prominently, arguing that 'For cyberspace to remain open and free, the same norms, principles and values that the EU upholds offline, should also apply online' (European Union 2013, p. 2). The relationship between the EU and the private sector is also seen as a normative one, and the strategy encourages the private sector to develop 'technical norms' that further interoperability between national communications networks (European Union 2013, p. 13). The strategy also casts the EU as a norm entrepreneur with an activist role in promoting particular norms internationally, saying the EU should 'encourage efforts to develop norms of behaviour' in a way that is consistent with its values, including 'human dignity, freedom, democracy, equality, the rule of law and the respect for

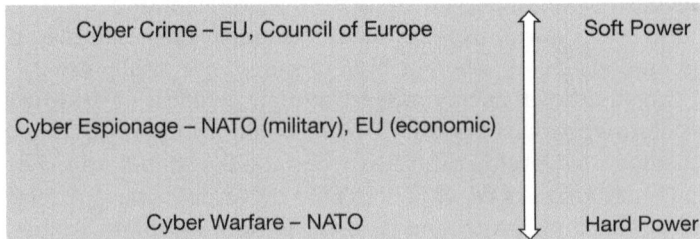

Figure 13.1 NATO and the EU – cyber complementarity.

fundamental rights' (European Union 2013, p. 15). The strategy seeks to support the development of 'norms of behaviour in cyberspace that all stake-holders should adhere to' and argues that 'Just as the EU expects citizens to respect civic duties, social responsibilities and laws online, so should states abide by norms and existing laws' (European Union 2013, p. 15).

Where the strategy really has a normative impact, though, is within the EU membership and in the area of information sharing and capacity building. To achieve greater resilience against cyber-attacks, the strategy requires providers of electronic communications services, including Internet Service Providers (ISPs), to report significant data breaches. The strategy also requires all EU member states to designate 'competent national authorities' to deal with Network Information Security (NIS), establish national Computer Emergency Response Teams (CERTs), and adopt an NIS national strategy and national cooperation plan to ensure appropriate EU-wide cooperation on cross-border cyber security threats. Additionally, providers of critical infrastructure across the European Union, including banks, telecommunications providers, the energy sector and transport operators, must 'assess the cybersecurity risks they face, ensure networks and information systems are reliable and resilient via appropriate risk management and share the identified information with the national NIS competent authorities' (European Union 2013, p. 6).

This far-reaching role for the European Union reflects the EU's strong nor-mative capacity as a regulator of private sector activity within the European region. However, there has been considerable resistance from the private sec-tor and a reluctance to abide by norms that entail significant financial costs to corporations. As Eneken Tikk argues:

> Private-sector organisations worry that disclosure of cyber attacks against them, and their results, might reduce trust in their business model or services. But government responses to politically motivated cyber attacks often require publication of such information. A balance needs to be struck between these public and private-sector interests (Tikk 2011, p. 127).

'Buying in' to norms is thus a conceptual and actual problem, as alluded to earlier in the chapter. Nevertheless, the strategy argues that the 'take up of a cybersecurity culture could enhance business opportunities and competi-tiveness in the private sector, which could make cybersecurity a selling point' (European Union 2013, p. 6). Companies are increasingly acknowledging the reputational costs of cyber-attacks and this may encourage norm adoption in that sector. Sony Pictures stock price dropped by 10 per cent in the wake of the Sony hack in 2014, for example, largely due to perceptions that the com-pany had lax cyber security standards (La Monica, 2014).

Running in tandem to the EU's efforts, the Council of Europe's Convention on Cybercrime (the Budapest Convention), passed in June 2001, is perhaps 'the most far-reaching multilateral agreement on cybercrime in existence'

(Fahey 2014). Ratified by 30 states, including the majority of EU members and Australia, Japan and the United States, the treaty is based on encouraging international cooperation on cybercrime through investigating, prosecuting and enhancing extradition procedures. Signatories to the convention agree to provide assistance to other countries when asked to investigate cybercrime, agree to pass laws that make certain activities criminal offences, such as hacking into computer systems, and commit to updating existing laws to reflect recent advances in ICT. Arguably the convention serves as a guideline for other countries wanting to advance their legal frameworks around cyber security (Marion 2010, p. 706) and is understood to be a legal framework to develop norms both internally and externally of the European region (Fahey 2014, p. 47).

Cyber security norm diffusion?

While there has been very real progress in the establishment of cyber security norms in the North Atlantic and European area, global acceptance of the types of normative standards and expectations that have been put in place do not automatically translate to other regional or cultural contexts. That only a small number of non 'Western' states have ratified the Budapest Convention speaks to these obstacles. The international architecture that has been built after the Second World War has established a rules-based order, but as new technologies reshape behaviour it should not be expected that the developing world, or indeed the emerging powerhouses of global politics, including the BRICS countries, will abide by norms that are based on preserving Western interests in cyberspace. Amitav Acharya's conception of the process of norm diffusion, mentioned briefly earlier in the chapter, is particularly relevant here. Norms may indeed spread from 'the west' to the 'east', or 'global south', but will likely be subject to processes of reinterpretation and reformulation to fit the particular cultural, social, economic and political needs of the 'norm-taker'. In other words, norms are likely to be localised as they are being diffused.

What might cyber security norms look like in a non-Western context? That is difficult to determine at this stage, especially as regional organisations representing other world regions, such as ASEAN and APEC, are behind in the extent and strength of regional cooperation on cyber security issues. The lack of internet penetration in the developing world is also seen as more of a problem than cyber security per se – only 2.5 per cent of people in Myanmar, for example, have access to the internet (Internet Live Stats 2016).

Additionally, as previously alluded to, many countries do not accept that the internet should be free of governmental control. Russia and China take this position but so do many other states. The authorities of Malaysia, for example, are deeply concerned about the social and cultural effects of online gambling and pornography and are implementing tighter restrictions on internet use in response to recent political controversies (Freedom House 2015).

This suggests that cyber security norms in other regions will be based on varying degrees of internet censorship.

Norm diffusion also requires norm entrepreneurship to extend beyond the original area of advocacy. Norms are unlikely to be adopted by states if they see no benefit in them. In that respect norms have to be attractive to a norm-taker. But it is also true to say that without persuasion, the other key element of soft power, norm cascades are less likely to occur. In this respect, both NATO and the EU have established channels of communication and persuasion that are already being utilised to build support for cyber security norms in states outside of their memberships. NATO's main mechanism for norm diffusion are its partnership agreements. These now encompass a wide range of states through the Mediterranean Dialogue, the Istanbul Cooperation Initiative and the Partners Across the Globe group of countries. Many of the partnership agreements with these countries include provisions for cyber security coopera-tion and there is significant appetite in countries like Japan, Mongolia, South Korea, Australia and New Zealand, for example, for cooperation with NATO on cyber security and to benefit more broadly from the norms that NATO is promoting in cyberspace (NATO Global Perceptions Project 2015).

The EU too has a strong foundation for promoting cyber security out-side of the European area, not least through the European External Action Service, and through the EU's strategic partnerships and relations with other regional organisations. The EU's newly released Global Strategy for the European Union's Foreign and Security Policy suggests that progress on cyber security governance depends on 'a progressive alliance between states, inter-national organisations, industry, civil society and technical experts' and this basic premise is reflected in the 2013 cyber security strategy too (European Union 2016, p. 43). EU norm entrepreneurship outside of its region continues through the process of 'mainstreaming' cyber security issues into EU exter-nal relations and the Common Foreign and Security Policy, and increasing engagement with key international partners. Most recently, the EU has estab-lished priority areas for cyber dialogues with countries outside the European Union area, which include protecting the digital economy, reducing cyber-crime, enhancing international stability, protecting the free and open internet and capacity-building in third countries. Finally, the Council of the European Union has established common principles for global cyber diplomacy with third countries, including a commitment to promoting 'norms for responsible state behaviour in cyberspace' and the application of existing international law to cyberspace (Council of the European Union 2015, p. 7). Because norms are socially constructed through dialogue and diplomacy, these activities are likely to have at least some influence on norm diffusion.

Conclusion

It is too soon to determine whether the measures taken by the EU and NATO will have a lasting normative impact. Overall, however, the various mechanisms for cyber security cooperation described in this chapter have provided European and North American states foundational tools with which to confront cyber security threats and challenges. Do they constitute a norm cascade? Probably a regional/internal one at least. Have they been fully internalised? The adoption of the EU and NATO cyber security strategies suggests a widespread acceptance within the Euro-Atlantic area that there are certain expectations of behaviour in the cyber security sphere and legal norms that should govern state and private sector behaviour. Collectively, the measures put in place by the EU and NATO constitute the most far-reaching normative frameworks currently in existence and complement the work done in other international organisations, including the G20, UN, OSCE and OAS.

At the global level, cyber security norms are still at an early stage of development and much work needs to be done in academia, policy making and within international organisations to establish what universally accepted expectations of behaviour might look like. There are some serious differences within the international society of states on how internet governance should work, and the incentives for states and non-state actors to develop cybercrime, cyber espionage and cyber warfare capabilities will continue to be present for many years. Nevertheless, norms are beginning to take root in the international system because of sustained norm entrepreneurship by regional organisations like NATO and the EU and there are at least some elements of these emerging normative frameworks which are likely to be of interest and acceptable across different cultural and geographic contexts. Welcoming norm entrepreneurship from non-Western states, understanding how norms are likely to be diffused and localised and respecting differences in how the role of the internet is perceived, will likely be the most important determinants of the success of the EU's and NATO's global cyber diplomacy.

References

Acharya, A. 2013, 'R2P and Norm Diffusion: Towards a Framework of Norm Circulation', *The Global Responsibility to Protect*, 5(4): 466–479.

Austin, G. 2016, 'International Legal Norms in Cyberspace: Evolution of China's National Security Motivations', In A. Osulaand, H. Rõigas (eds), *International Cyber Norms: Legal, Policy & Industry Perspectives*. Tallinn: NATO CCD COE Publications, 177–201.

Barnes, J. E. 2016, 'NATO Recognizes Cyberspace as New Frontier in Defense', *Wall Street Journal*, viewed 11 May 2017, www.wsj.com/articles/nato-to-recognize-cyberspace-as-new-frontier-in-defense-1465908566.

Chaban, N. et al. 2015, 'Perceptions of 'Normative Power Europe' in the Shadow of the Eurozone Debt Crisis: Public Perspectives on European Integration from the Asia Pacific', In A. Björkdahl, N. Chaban, J. Leslie, A. Masselot (eds), *Importing EU Norms, Conceptual Framework and Empirical Findings*. Heidelberg: Springer, 55–77.

Clark, C. 2016, 'NATO Declares Cyber A Domain; NATO SecGen Waves Off Trump', *Breaking Defence*, viewed 11 May 2017, http://breakingdefense.com/2016/06/nato-declares-cyber-a-domain-nato-secgen-waves-off-trump/.

Council of the European Union. 2015, 'Council Conclusions on Cyber Diplomacy', viewed 11 May 2017, http://data.consilium.europa.eu/doc/document/ST-6122-2015-INIT/en/pdf.

Denning, D. E. 1999, 'Activism, Hacktivism, and Cyberterrorism: The Internet as a Tool for Influencing Foreign Policy', *Global Problem Solving Information Technology and Tools*, viewed 11 May 2017, http://nautilus.org/global-problem-solving/activism-hacktivism-and-cyberterrorism-the-internet-as-a-tool-for-influencing-foreign-policy-2/.

European Union. 2013, 'Cybersecurity Strategy of the European Union: An Open, Safe and Secure Cyberspace', viewed 11 May 2017, https://eeas.europa.eu/policies/eu-cyber-security/cybsec_comm_en.pdf.

European Union. 2016, 'Global Strategy for the European Union's Foreign and Security', viewed 11 May 2017, https://europa.eu/globalstrategy/en/global-strategy-foreign-and-security-policy-european-union.

Fahey, E. 2014, 'The EU's Cybercrime and Cyber-Security Rulemaking: Mapping the Internal and External Dimensions of EU Security', *European Journal of Risk Regulation*, 5(1): 46–60.

Fiddner, D. 2018, 'Cyberspace's Ontological Implications for National Security', In A. Gruszczak and P. Frankowski (eds), *Technology, Ethics and the Protocols of Modern War*, London: Routledge (forthcoming).

Finnemore, M. and Sikkink K. 1998, 'International Norm Dynamics and Political Change', *International Organization*, 52(4): 887–917.

Freedom House. 2015, 'Freedom on the Net, Malaysia', viewed 11 May 2017, https://freedomhouse.org/report/freedom-net/2015/malaysia.

Internet Live Stats. 2016, 'Elaboration of data by International Telecommunication Union (ITU)', *United Nations Population Division*, viewed 11 May 2017, www.InternetLiveStats.com.

La Monica, P. R. 2014, 'Sony Hack Sends Stock Down 10% in Past Week', *CNN*, viewed 11 May 2017, http://money.cnn.com/2014/12/15/investing/sony-stock-hack/.

Lewis, J. A. 2014, 'Liberty, Equality, Connectivity: Transatlantic Cybersecurity Norms', *Centerfor Strategic and International Studies*, viewed 11 May 2017, www.ciaonet.org/attachments/24812/uploads.

Marion, N. 2010, 'The Council of Europe's Cyber Crime Treaty: An Exercise in Symbolic Legislation, *International Journal of Cyber Criminology*, 4(1/2): 699–712.

NATO. 2010, 'Strategic Concept Strategic Concept 'Active Engagement, Modern Defence'', viewed 11 May 2017, www.nato.int/cps/en/natohq/topics_82705.htm.

NATO Global Perceptions Project. 2015, 'Interviews in Japan, Mongolia, South Korea, Australia and New Zealand', viewed 11 May 2017, www.ttu.ee/projects/nato-global-perceptions/.

Prime Minister of Australia. 2017, 'Australia and China Agree to Cooperate on Cyber Security', viewed 11 May 2017, https://www.pm.gov.au/media/australia-and-china-agree-cooperate-cyber-security.

Rõigas, H. 2015, 'The Ukraine Crisis as a Test for Proposed Cyber Norms', In K. Geers (ed), *Cyber War in Perspective: Russian Aggression against Ukraine*. Tallinn: Cooperative Cyber Defence Centre of Excellence.

Schmitt, M. N. (ed). 2013, *Tallinn Manual on the International Law Applicable to Cyber Warfare*. Cambridge: Cambridge University Press.

Sunstein, C. R. 1996. 'Social Norms and Social Roles', *Columbia Law Review*, 96(4): 903–968.

Tikk, E. 2011, 'Ten Rules for Cyber Security', *Survival*, 53(3): 119–132.

Valeriano, B. and Maness, R. C. 2015, *Cyber War versus Cyber Realities: Cyber Conflict in the International System*. Oxford: Oxford University Press.

Vihul, L. 2017, 'Tallinn Manual 2.0 Clarifies How International Law Applies to Cyber Operations', viewed 11 May 2017, hwww.atlanticcouncil.org/news/press-releases/tallinn-manual-2-0-clarifies-how-international-law-applies-to-cyber-operations.

White House. 2011, 'International Strategy for Cyberspace: Prosperity, Security, and Openness in a Networked World', viewed 11 May 2017, https://obamawhitehouse.archives.gov/the-press-office/2015/09/25/fact-sheet-president-xi-jinpings-state-visit-united-states.

White House. 2015, 'Fact Sheet: Xi Jinping's State Visit to the United States', viewed 11 May 2017, www.whitehouse.gov/the-press-office/2015/09/25/fact-sheet-president-xi-jinpings-state-visit-united-states.

Williams, K. B. 2015, 'G20 Nations Reach Anti-Hacking Pledge', viewed 11 May 2017, http://thehill.com/policy/cybersecurity/260414-g20-nations-reach-anti-hacking-pledge.

14 Cyberspace's ontological implications for national security

Dighton (Mac) Fiddner[1]

Introduction

Having a cyberspace national security policy, and the strategy to implement that policy, is signally important because of the increasing prevalence of cyber incidents from nation states, nongovernmental organisations (including organised crime and terrorist groups) and skilled individuals that can have a significant impact on a nation's ability to defend itself and continue its very existence. Advanced industrial and post-industrial states have become so dependent on power, communications, transportation and financial infrastructures that even short disruptions would have significant effects (Choucri 2012, p. 149). Moreover, it is widely recognised that the infrastructures are so interlinked and mutually dependent that any serious disruption is likely to have cascading effects,[2] eventually affecting the power grid on which the others depend for their continued operation (Williams 2011).

The risk acceptance of critical infrastructures should be identified and prioritised in order to determine what constitutes a national security risk instead of a 'nuisance' (cyberactivity that is annoying or interferes with the operation of the infrastructure but does not rise to the level of threatening the national cyber-infrastructure [or one or more of the other national critical infrastructures or significant portions of them] to no longer be able to function as a system).

Even with the potential for serious deleterious damage to the health, safety, economy and public confidence of society[3] in the (most-) developed states, cyberspace national security policy and strategy have been poorly conceptualised, remains underdeveloped and, while a great deal of innovative thinking about cyber issues has taken place, cyberspace national security has been more difficult to crystallise consensus around than the essentials of nuclear security (Williams 2011).

Some of this lack of national-level consensus on how best to protect U.S. national security might be attributed to the difficulty of conceptualising and understanding the medium of cyberspace (its ontology). In many ways, it is different from the other mediums that could threaten national security. Cyberspace is a strategic domain characterised by stealth, plausible deniability, accusation rather than proof and suspicion rather than knowledge that can create and/or attenuate uncertainty for decision makers. It is both a global commons and a strategic domain inextricably linked to the other physical strategic domains. It is an asset and liability, source of strength and weakness and facilitator of both repression and revolution. It is ubiquitous and elusive, characterised by paradox and complexity. It is also a diffuse medium rather than a concentration of destructive power that has developed over time and is symbolised by Facebook and Twitter.

Nazli Choucri in *Cyberpolitics in International Relations* informs us that 'the social science vocabulary for this new arena is at an early stage of development. The technical terminology is being created as technological parameters are expanded. But, there is no agreed-upon ontology for all facets of cyberspace' (Choucri 2012, p. 240).

> Some of the critical and determining contentions surrounding cyberpolitics are about defining the domain itself. [...] If the issue area is not well defined, uncertainty and ambiguity often reflect an emerging 'marker for loyalties' and [...] if the contours of an issue are obscure, then clarifying meaning and value becomes the core of the contention. Far from being a matter of semantics, this definitional phase is fundamental in shaping the playing field in terms of power and influence (Choucri 2012, pp. 250, 253).

Or, as David Nachmias and Chava Frankfort-Nachmias inform us, 'if concepts are to serve as the basis for empirical research they must be: clear, precise, and unambiguous, and they must mean the same thing for all of the people – researchers, their subjects, and those who will "consume" the research' (Frankfort-Nachmias and Nachmias 2008).

The term 'cyberspace' today means many different things to different people. For 'technologists' cyberspace is essentially a technical transmission system; others envision it as an ecosystem (Herrera-Flanigan 2001; Department of Homeland Security 2011), an ecology (Air Force Research Laboratory 2003), a combination of the two (Baudoin 2013) or a layered system, as Martin Libicki (2009) and David Clark (2010) describe it. Additional knowledge of the nature of cyberspace should lead to better understanding of what is happening strategically and to what, how and when to respond and with what, if warranted.

In cyberspace's case, decision making and strategy transcend the purely technical realm and incorporate the same conditions as all other national security issues, necessitating broader solutions rather than purely technical ones. Cyberspace is a system comprised of hardware, digital data and human beings. It is a man-made system with inherent vulnerabilities as a result of its design and construction, especially the unique characteristics of the scale-free network structure. Cyberspace exists to facilitate human activity and is subject to human decision making with all of its foibles. Human decision making is not algorithmic but is based on different individuals' sometimes idiosyncratic preferences of costs and benefits, fundamental issues of politics and strategy, bargaining and escalation dynamics and control. Similarly, response decisions should be left to human decision makers and not predicated on some algorithmic response since the initial activity prompting a response is a human activity based on a human decision.

Therefore, since decision making is a human activity there should be a definitional bias toward an aggregate of individuals, organisations and the three interrelated physical, informational and cognitive dimensions that collect, process, disseminate or act on information.

Cyberspace strategic domain: Structure of cyberspace's strategic domain

The strategist community has traditionally postulated that activity, concern or function could occur in four separate, independent domains: land, sea, air and space, but cyberspace provides a fifth sphere where activity, concern or function can occur. Cyberspace is now generally accepted as being a fifth separate, independent strategic domain as well as a dimension of national power and instrument of that power (according to DoD Joint Publication 3–13, see U.S. Chairman of the Joint Chiefs of Staff 2012), but structured and operating differently than the traditional four. It has been generally accepted that even though each of the four traditional strategic domains can serve as the locus for activity, concern or function independently, those loci could catalytically (if conditions permit) create activity, concern or function in the other domains (i.e. what begins in one could also transfer to another or others so that the activity and its effects are occurring in multiple domains). Cyberspace is envisioned to exhibit that property also. Cyberspace could then be defined as:

> A man-made global strategic domain, dimension of national power, and instrument of the dimension of national power within the information environment consisting of the interdependent network of information technology infrastructures and resident data, including the Internet, telecommunications networks, computer systems, and embedded processors and controllers for the production and use of information by individuals and organisations (Fiddner 2015, p. 6).

However, cyberspace as a strategic domain has three unique properties (See Figure 14.1 for a visual representation of the five strategic domains):

1 It has no physical boundaries as the other domains do, which means cyberspace permeates the entire strategic environment.
2 It superposes (simultaneously occupies the same space) the other four domains.
3 It can generate activity without human initiation, i.e. digitally self-generated activity as a dimension of and instrument of national power.

This means that activity solely in the cyberspace domain can be constrained exclusively in the cyberspace domain, could transfer to one or more of the

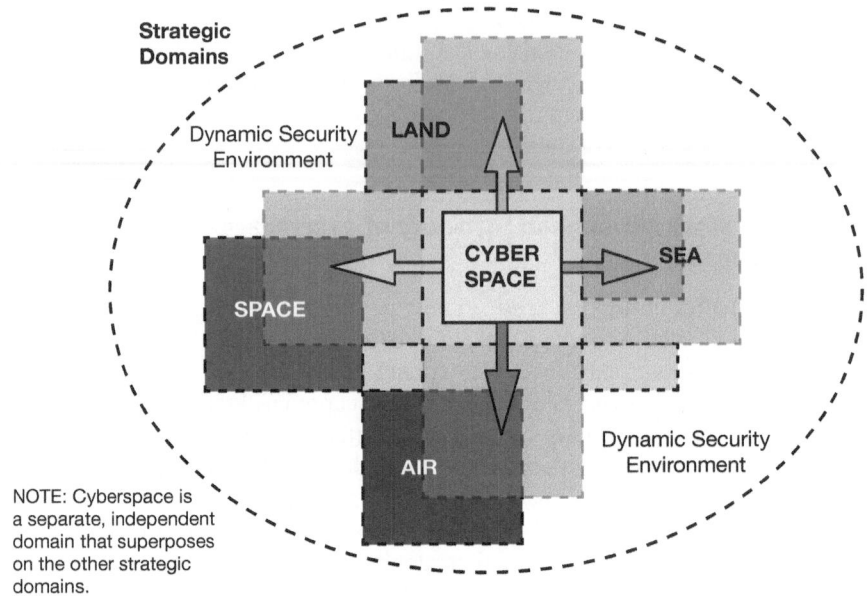

Figure 14.1 Strategic environment.

other four traditional domains or could simultaneously affect activity in one or more of the other domains without human initiation.

Much like the space strategic domain, cyberspace is a global common; national sovereignty over parts of or the entire domain does not exist, but sovereign tangible components (or instruments of national power) can exist within the domain. Further, unlike the other strategic domains, cyberspace superposes[4] the other strategic domains, i.e. it simultaneously occupies the space occupied by the other strategic domains so that the two coincide throughout their whole extent. As a consequence, activity within cyberspace can have a direct integral (not merely catalytic) causal effect on activity in the other strategic domains.[5]

Cyberspace simultaneously also has the characteristics of both a dimension of national power and an instrument of national power. As a dimension of national power, a nation's relative cyberspace power can be used in the very same ways as any other dimension of national power, either persuasively or coercively to entice or deter/compel an entity to do what it might otherwise wish not to do or not do what it otherwise might wish to do. At the same time, the components (interdependent network of information technology infrastructures and resident data) could conceptually be viewed as cyber instruments of national power.[6] Employment of these cyber instruments would be an application of cyberspace's dimension of national power. This conceptualisation of cyber instruments is analogous to examples of the other dimensions of national power as displayed in Figure 14.2 (Table 14.1).

Table 14.1 Dimensions and instruments of national power

Dimensions of national power	Examples of instruments of national power				
Economic	Trade regulations	Tax and monetary policy	Foreign aid	Economic sanctions	
Political	Diplomats	International organisations	International laws	Norms	Political sanctions
Military	Conventional forces	Military assistance	Weapons of mass destruction	Special ops	
Cyber	Malware	Restriction of services	Social media monitoring & use	Monitoring & collecting transmissions	Physical destruction of controlled processes (Stuxnet)

Source: Chart adapted (with additions by the author) from similar handout to students from Department of National Security and Strategy, U.S. Army War College, Carlisle, PA.

In other words, it is almost as if cyberspace is an *uber*-strategic domain. However, since cyberspace is man-made, and is already in place, the implication for national decision makers is that what exists is what they have to work with. Certain properties or characteristics were deliberately designed, to include its network topography, which inherently puts the system's network(s) (or cyberspace) at risk.

The cyberinfrastructure system is a scale-free instead of an exponential network. In contrast to an exponential, scale-free network connectivity is structured through a few highly connected nodes while most of the nodes are just 'end users' not connected to other nodes; the removal of 'end user' nodes does not affect the paths between the remaining network's nodes. Under a **targeted** deletion attack, when highly connected nodes are eliminated, the paths of the remaining nodes to other nodes are greatly diminished. And, if a vital (or a highly connected) node is deleted, catastrophic fragmentation of the scale-free network into many isolated clusters is swift. Any informed agent that attempts to deliberately damage a cyberinfrastructure will not do so randomly, but will preferentially **target** in descending order of connectedness, or some other criteria of importance, to jeopardise the system most effectively. Consequently, defending such a system with its other vulnerabilities is extremely difficult and, strategically, national security decision makers must at least consider using dimensions of national power offensively when responding to cyberactivity.

Threat, response and spheres of interaction

Cyberspace now seems to be having a changing effect on security's threat and response vectors. The following diagram's (Figure 14.2) horizontal rows represent the sphere of interaction (cyber, merged or physical) in which activity at differing levels (individual, national and global) of interaction (the diagram's vertical columns) occurs. It seems to be diffusing the sphere from where threats emanate to jeopardise any level of security (personal, national and collective) and the possible response to those threats. Normally, the threat to security emanates from the same sphere of interaction in which the level of security is located, e.g. global cyber security is expected to be jeopardised by a threat from the cyber sphere and so on for each of the levels of security in their respective sphere of interaction. Conversely, in the traditional physical and cyber spheres of interaction, the response to a threat is most likely within the same sphere from which the threat emanated. The cases listed

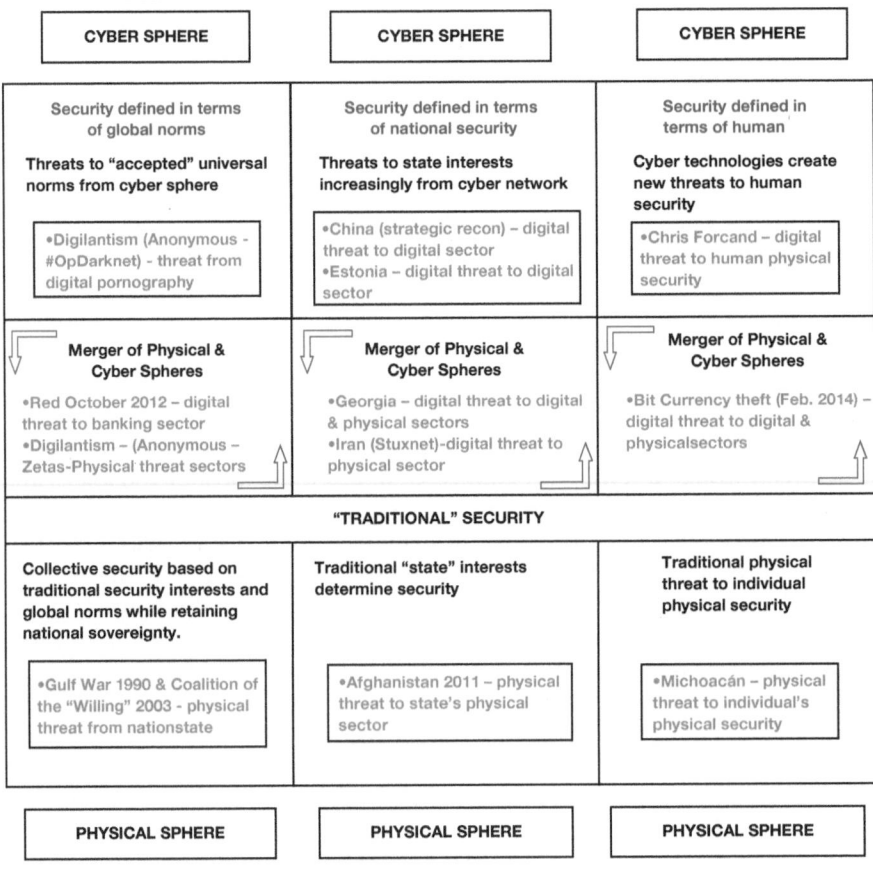

Figure 14.2 Threat and response vectors.

are representative of threats to the different levels of security to demonstrate possible threat origins and paths (vector). The diagram (and discussion following) is not intended to imply that cyberspace is not involved in actions within the more traditional physical sphere of interaction, because it most assuredly is; cyberspace is just too expansive and pervades all spheres of interaction (as well as all of the strategic domains through superposition), but in all cases, there is always the possibility of a physical threat to the cyberinfrastructure.

Cyber sphere of interaction

Global cyber sphere (left top square)

#OpDarknet (October 2011)

In the upper left square's case, the threat to security (pornographic images and videos) was from the global cyber sphere on the Hidden Wiki (located on the Tor Network's Hidden Service Protocol) from the underground paedophile community 'Lolita City'. Such images are considered a breach of the global norm against child pornography. Kelsy Ida has found that as cybertechnologies create global communities of like interests, they are taking responsibility to enforce generally accepted norms of social behaviour. Cyber vigilantes ('Digilantes') 'have the ability to rationally recognize the problems of the status quo, and also, perhaps, the power to redress the situation […] and […] "act as collaborative regulators," particularly in the absence of "a single coercive actor with Weberian 'legitimate monopoly on the use of force'"' (Ida 2016, p. 524).

Anonymous (a global cyber collective), in early October 2011, responded in the cyber sphere of interaction initially by removing the links. Shortly after the links returned online, the Lolita City site became inaccessible in its entirety, presumably as a result of DDoS attacks initiated by Anonymous. Following the initial response, Anonymous members followed up their response after discovering nearly all of the pornography's digital fingerprint and issuing the host a warning to remove the content from its server at 9 p.m. (CST) on 14 October. Freedom Hosting refused to comply and two and a half hours later, Anonymous completely shut down Freedom's services with DDoS attacks that created a 1GB SQL and 100,000 ASCII files of Guy Fawkes masks every five minutes.

On 18 October, Anonymous released the names of the 1589 users of Lolita City via PasteBin, including their username, volume of images uploaded, and age of the account. Moving their response to the physical sphere of interaction, Anonymous invited Interpol and the FBI to investigate the records further. 'We vowed to fight for the defenseless, there is none more defenseless than innocent children being exploited' (Know Your Meme 2014). The FBI responded in the cyber sphere of interaction by installing some malware that infected all users who accessed the '. onion' sites during the occupation period which unmasked the TOR routing protocol and revealed the users' real locations (Know Your Meme 2014).

National cyber sphere (centre top square)

China

As demonstrated by the Chinese case, the cyber sphere threat can be combined with threats from the physical sphere as well if needed. The Chinese aggressively probe and enter global cyber sphere networks to gain a competitive advantage in economic matters, business, military and political bargaining (Mandiant 2013), but also for truly strategic reasons: to win victory before the first battle ... by mapping the opponent's digital terrain' (Thomas 2016, p. 174). Most of their behaviour is driven by the belief that the US maintains hegemonic power over global cyberspace, that information superiority is becoming a key component of national power and, therefore, they are at a strategic disadvantage in any conflict with the US (and their allies).

Strategic digital reconnaissance comes from activities specifically targeted and directed for such purposes to provide knowledge of the digital 'landscape', or virtual *shi,* to allow more effective offensive and defensive activities, if needed. Active offense (system sabotage) is the Chinese preferred strategy for winning a cyber conflict. In such an offensive move, the Chinese will seek to damage or disrupt the opponent's cyber system's material and technical foundational critical nodes making it impossible to adjust to problems on the battlefield. By controlling information, the opponent is essentially left in the dark about what is going on and is hindered and limited to what it can do, making it impossible to turn war potential into actual capabilities for engaging in war.

Chinese strategic thought does not foresee information deterrence acting alone; 'informatized warfare can increase its deterrent power capable of achieving strategic objectives when combined with nuclear deterrence capabilities' (Thomas 2016, p. 196). Along with the nuclear deterrent, China foresees conventional deterrence, space deterrence and information deterrence as a 'cocktail' for use in future conflicts (Thomas 2016, pp. 187–189).

Estonia

Much like China, Russia views cyberspace strategically as asymmetric where cyber war, economic sanctions, a domestic and international public information campaign, the manipulation of youth organisations or gangs and the penetration of key sectors of the economy and subversion of politicians become a surrogate for large-scale military capabilities that are unavailable or simply not usable. The target's patriotic hackers will, probably along with vital socio-political institutional structures, become early and primary targets of future cyberspace offensive activity to deprive states of their ability to retaliate. The Russian experience in both Estonia and in Georgia indicate that Moscow operationalised strategic information war to achieve victory by paralysing the targeted country's social infrastructure networks, i.e. what might be called its central nervous system (Blank 2016).

In Estonia, Russia[7] appears to have employed this strategic concept with threats from the cyber sphere of interaction to attack Estonia's cyberinfrastructure (one of the world's most 'connected' governments and societies at the time) and jeopardise that state's ability to function and retaliate in the cyber sphere of interaction. The attacks on Estonian socio-economic and political institutions were allegedly coordinated with organised crime structures like the Russian Business Network[8] and were a reaction to Estonian authorities' transferal of the Bronze Soldier of Soviet Liberators of Estonia from the Nazis monument in Tallinn to another site. This 'war' was also combined with attempts to incite domestic violence in Estonia, attack its embassy in Moscow through violent demonstrations orchestrated by Nashi (one of the 'official' Russian youth organisations), an ongoing public diplomacy campaign targeting both domestic Russian and Western audiences to label Estonia's regime as Fascistic and equally ongoing economic warfare (Blank 2016).

The cyber-attacks appeared to have been long-planned to disrupt and possibly unhinge the Estonian government and society and to demonstrate NATO's incapacity for protecting Estonia against this novel form of attack. Undoubtedly as well, this operation aimed to compel Estonia to take Russian interests into account in its policies and, even though it was a bloodless and non-violent attack, also to punish the monument transferal and deter others from following suit by holding it up as an example of the risks to anyone who crosses Russia (Blank 2016). Estonia's response was mostly defensive in the cyber sphere of interaction due to uncertainty in definitive attribution of the attacks.

Individual cyber sphere (right top square)

Anonymous – Chris Forcand

The threat (and response) in the Chris Forcand case is once again completely in the cyber sphere of interaction involving a breach of the generally accepted universal norm against child pornography. Only later did Anonymous' response with publication of the cyber data move from the cyber sphere of interaction to the physical sphere of interaction and Forcand's arrest by authorities. The arrest of Chris Forcand illustrates one of the first examples of cybervigilantism for individual, personal physical security or safety.

On 7 December 2007, alleged Internet predator Chris Forcand, 53, was charged with two counts of luring a child under the age of 14, attempting to invite sexual touching, attempted exposure, possessing a dangerous weapon and carrying a concealed weapon. Forcand was tracked by cyber-vigilantes who 'seek to out anyone who presents a sexual interest in children'. Members of Anonymous 'contacted the police after some members were "propositioned" by Forcand with "disgusting photos of himself"' (Crumrine 2008). Sexually explicit conversations were then forwarded to Forcand's church and a blog he wrote at praize.com and his name, address and phone number were posted online (Jenkins 2007). The report also stated that this is the first time a suspected Internet predator was arrested by the police as a result of Internet vigilantism (Wikipedia 2014).

Physical sphere of interaction (traditional 'security')

Global-physical sphere (left bottom square)

The global-physical sphere is not an unusual arena of action within international security. States collaborate ostensibly to enforce an accepted global norm. Although primarily physical, instruments of the cyber dimensions of national power are increasingly being used in conjunction with instruments of the military and other dimensions of national power.

Gulf War 1990

A coalition of 34 countries opposing Iraq's aggression and violation of the global norm of Kuwait's universally recognised territorial sovereignty was formed under the auspices of the United Nations to punish Saddam Hussein and to restore Kuwait's internationally recognised border. Cyber activity was used to augment the physical instruments of collective national power to provide an even greater comparative advantage.

U.S. Coalition of the Willing 2003

The US's 'coalition of the willing' represents an example of states collaborating to gain compliance with a generally agreed upon global norm, i.e. Saddam Hussein's non-compliance with UN resolutions regarding his program of weapons of mass destruction. In 2003, even without the endorsement of the UN Security Council, the United States decided, based on its intelligence data, that Saddam's program of trying to acquire or to develop weapons of mass destruction was a violation of numerous UN Security Council resolutions designed to prevent him from doing so. They subsequently formed a coalition of willing partners to punish Saddam and, ultimately, to intervene militarily, to bring him back into compliance with the resolutions. Although primarily within the physical sphere of interaction, activity within the cyber sphere of interaction reinforced the coalition's comparative advantage of the physical sphere of interaction's national instruments of power.

National physical sphere (centre bottom square)

This represents the usual realist notion of national security; a state acting within the physical sphere of interaction for no other reason than its own self-interest. The US retaliation against Afghanistan for providing sanctuary to al Qaeda and refusing to turn al Qaeda's leaders over to US authorities is an example of behaviour in this arena of action. Activity within the cyber sphere of interaction greatly enhanced both the initial deployment of the US's physical sphere of interaction's instruments of military power and the subsequent larger scale intervention by the US and the international coalition supporting it.

Individual physical sphere (right bottom square)

Michoacán

Humans living in locales desiring physical safety and the basic necessities of every-day life seem to be turning to anyone who can provide for them. The situation in the Mexican state of Michoacán seems to substantiate this observation. The threat was in the physical sphere of interaction and the response was there as well; this is not to suggest that there was no cyber sphere of interaction involved but only that the principal threat and response were both in the physical sphere of interaction.

When authorities there could or would not provide safety for the populace (physical sphere of interaction), self-defence groups (vigilantes) emerged hoping to drive the Knights Templar drug cartel (which ran an extensive extortion racket and had come to control a number of local governments, as well as much of the agricultural business in the region) out of Michoacán. In late January 2014, the Mexican federal government sent a surge of troops and federal police to the region (physical sphere of interaction) after the vigilantes began seizing control of communities around a key Knights Templar stronghold and openly declaring their intention to attack the cartel members there. The situation calmed with the arrival of the troops and officers, who controlled 27 of Michoacán's 113 municipalities. Though federal authorities demanded the vigilantes lay down their arms, they continued to sport assault rifles and other weapons at road blocks outside the towns they had seized. There were some early standoffs between government forces and vigilante groups over the demand that they disarm, but they later appeared to be cooperating in some parts of the state (Fausset 2014).

Merged physical-cyber sphere

Global merged physical-cyber sphere (middle left square)

Red October (October 2012)

There is no compelling case in this category of threat/response vectors that provides a clear example of a threat in one sphere of interaction and a response in another. However, the threat in this case was from the cyber sphere of interaction towards the banks' cyberinfrastructure (cyber sphere of interaction), but the true target was those banks' financial assets (physical sphere of interaction). As far as can be determined, there was no response but presumably if there was it would have been in the cyber sphere of interaction.

The Russian cybersecurity firm Kaspersky Labs discovered a worldwide cyber-attack dubbed 'Red October' in October 2012 that had been operating since at least 2007. The campaign appeared at the least to be an example of strategic economic or political reconnaissance and espionage by some entity. The true identity of the perpetrators has not been definitively determined since the C&C architecture is arranged to hide the mothership-server through proxy functionality of every node in the malicious structure. The exploits appear to have been created by Chinese hackers, although many believe the perpetrators are the Russian Business Network[9] who are comfortable using Chinese malware and adapting it for their own use (The Hacker News 2013).

The primary targets of the attacks appeared to be countries in Eastern Europe, the former USSR and Central Asia, although Western Europe and North America reported victims as well. The virus collected information of the 'highest level' and included geopolitical data from governments, embassies, research firms, trade and commerce, aerospace, military installations, energy providers, nuclear and other critical infrastructures that could be used by nation-states or could be traded in the underground and sold to the highest bidder (NATO 2013; Holden 2013).

Zetas

In October 2011, following the kidnapping of an Anonymous member residing in the state of Veracruz, Anonymous threatened to publicise online the personal information of Los Zetas and their associates unless Los Zetas freed their kidnapped member by November 5th (Kan 2013a). Despite attempts at 'reverse hacking' and death threats sent to Anonymous members, Los Zetas released the kidnapped member on 4 November. Admittedly, they only did so with a warning to Anonymous that they would execute ten people for each name that Anonymous might publicise (Kan 2013a), but what is significant here is that Los Zetas 'blinked' first.

Notable here too is that Anonymous (through a local branch in Acuña) has since re-engaged Los Zetas, publishing photos of known cartel properties online with, thus far, little retribution. By choosing to 'out' the various parts of their 'organizational infrastructure', Anonymous has once more struck at Los Zetas' 'criminal brand'. However, here again the digilantes – with notable public support – have engaged organised crime even beyond state enforcement (Kan 2013b).

National merged physical-cyber sphere (middle centre square)

Georgia

Georgia represents a violation of the generally accepted universal norm of the inviolateness of a state's territorial integrity by another state's military instrument of power (physical sphere of interaction) while, at the same time, the use of the cyber sphere of interaction to enhance the military instrument of power directly and indirectly by disrupting government functions, social and media connections and economic activity. Georgia, for the first time, was an attempt to attack military forces' command and control and weapons systems (cyber sphere of interaction) on the one hand, and information-psychological attacks against media, communications and perceptions (cyber and physical spheres of interaction) on the other (Blank 2016). The organisers of the cyberactivity[10] were recruited through the Internet and social technology and aided by Russian organised crime, even to the point of having hosting software ready for use. They also probably had advance notice of Russian military intentions and were tipped off about the timing of Russian military operations while they were taking place (Blank 2016).[11]

The first wave of cyber-attacks on 6–7 August 2008, 24–48 hours before the actual war,[12] were carried out by botnets and C2 systems that were prepared before the invasion and associated with Russian organised crime. After this, the second wave resorted, though not exclusively, to postings containing both the cyber-attack tools and lists of suggested targets for attacks on websites (Blank 2016), a carryover from Estonia. Once Russian troops had established positions in Georgia, the attack list expanded to include many more government websites, financial institutions, business groups, educational institutions, and more news media websites and a Georgian hacking forum to preclude any effective or organised response to the Russian presence and induce uncertainty as to what Moscow's forces might do.

These attacks significantly degraded the Georgian government's ability to deal with the invasion by disrupting communications between it and Georgian society, stopping many financial transactions and causing widespread confusion. The clear objective of the cyber-strikes was to support and further the goals of the military operations and they were timed to be numerous within hours of the first Russian military operations and ended just after those operations ended. Beyond this, the cyber campaign was part of a larger information battle between Russian media and the Georgian and Western media for control of the narrative (Melikishvili quoted in Libicki 2016, p. 251).

Stuxnet

Stuxnet, the computer malware designed to attack Iran's nuclear facilities, was initially discovered in June 2010 and is generally attributed (although not confirmed) to the United States and Israel. The worm includes a highly specialised malware (cyber sphere of interaction) that is designed to target only Siemens supervisory control and data acquisition (SCADA) systems (cyber sphere of interaction) in Iranian nuclear enrichment centrifuges. It subsequently almost ruined one-fifth of the Iranian nuclear centrifuges (physical sphere of interaction) by causing them to spin out of control while simultaneously displaying the normal centrifuge values during the attack. 'The attackers took great care to make sure that only their designated targets were hit. It was a marksman's job' (Broad et al. 2011).

Individual merged physical-cyber sphere (middle right square)

Bit (Crypto) Currency 2013–14

Crypto currency thefts and subsequent near collapse at the end of 2013 and into 2014 seem to demonstrate a case where actions in cyberspace violated generally accepted norms of accountability and led to responses in both the cyber and physical spheres of interaction. The appeal of this crypto currency operating on the WWW's 'Deep Web' (DarkNet or Tor – The Onion Router) network was that it was 'outside' normal governance of regulated currency and economic commerce. Therefore, people were able to make one-to-one transactions to buy goods and services and exchange money across borders without involving banks, credit card issuers or other third parties.

Between the end of 2013 and into 2014, several exchanges saw their repository of legitimate legal tender used to purchase the crypto currency significantly reduced or depleted with an estimated loss of $623.22 million: Pony Botnet ($220,000), Mt. Gox ($500,000), Silk Road 2 ($2.7 million), Sheep Marketplace ($56.4 million), Silk Road ($127.4 million) and Mt. Gox ($436 million). Identifying those responsible for the losses exposed them to possible physical security or economic loss, more likely through vigilantism since the 'virtual' currency was purchased with tangible legal tender of different nation states. Most of these thefts appear to have been committed by 'insiders' responsible for maintaining or administering the exchanges. Users 'responded with anger (obviously) and threats. But Bitcoin being Bitcoin, the money was lost and gone forever' (comments of DefCon, administrator of Silk Road exchange, in Moktadir 2014). The facts and actions in this case are not finished, nor are they conclusive yet, but tend to suggest some physical action will be forthcoming from either individuals suffering losses (most likely) or possibly from a more traditional level of governance.[13] Autumn Radtke, the CEO of First Meta (an online currency exchange), died under mysterious circumstances at her home in Singapore as well as other bankers at JPMorgan, Deutsche Bank and Zurich Insurance Group (David 2014).

Conclusions

Response to and crisis management of cyber-incidents could prove much more problematic than during the Cold War, especially where definitive attribution of activity is mostly impossible if the perpetrator prefers anonymity. Without a solid conceptual foundation, a cyberconflict would be nearly impossible to navigate. Cyberspace is a more complex strategic domain than the other four strategic domains (air, land, sea and space) and, therefore, demands more reasoned response calculations for optimal decisions about cyberactivity, especially during a crisis.

The unique nature of cyberspace's strategic domain as both an independent space where cyberactivity takes place and one that superposes the other four strategic domains presages significant difficulty for the strategic planner and decision maker to identify accurately the true locus of the activity, attribution and intent of the perpetrator, time available to respond and, subsequently, response. Also, cyberspace's inherent three functions, i.e. global strategic domain, dimension of national power and an instrument of that dimension of national power, further complicates its use. Activity in the different functions will have different ramifications and decision makers need to consider specifically which of those three functions the observed cyberactivity is employing and which would be most effective as a response, if warranted. Only better knowledge of the cyber domain and its role and function in the strategic environment will allow decision makers to identify the different strategic options available and, consequently, lead to more sophisticated anticipation and better, more nuanced acceptable cost-benefit responses.

However, national security in cyberspace is not a narrow technical issue; it involves fundamental issues of politics and strategy, great power relations, bargaining and escalation dynamics and control, but an understanding of both the technological domain and strategic environment is imperative to making the optimal responses to actions against the nation and its cyber/critical infrastructure. Making decisions in the strategic context of cyberspace is as much about managing uncertainty as achieving a specific established goal. Successful strategic decision making and crisis management then is based upon clearly identifying goals, understanding profoundly the strategic environment and assessing one's comparative advantage within that environment, calculating costs through an objective appraisal of alternative strategies' resources and benefits – developing consistency between the internal environment and the external environment, between the goals and values, resources and capabilities, and structure and systems and range of options an understanding of cyberspace's domain provides.

Perhaps the best advice to senior national security decision makers will be to avoid premature doctrinal and organisational hardening in favour of teams of rivals and lots of options since the conditions, identified by Bernard Brodie in 1959, for nuclear strategy seem to obtain for cyberspace today:

> It is necessary to consider American national policy in the light of the facts that (a) today the defense in all its aspects is immeasurably behind the offense in effectiveness; (b) there is no present reason to suppose that future technological developments will, on balance, drastically favor the defense over the offense; and (c) the amount of damage that the United States must therefore expect to receive from [any Soviet nuclear] a strategic [cyberattack] will be intolerably huge
>
> (Brodie 1959, pp. 226–227).

Notes

1 The author thanks the IBM Center for The Business of Government for generously funding the foundational research which produced this article. An earlier version of this manuscript was presented at the CISS Interdisciplinary Approaches to Security in the Changing World Millennium Conference, Jagiellonian University, Krakow, Poland, 18–20 June 2015.

2 See Perrow (1984) for an extensive explanation of the dangers of highly complex, interconnected systems.

3 Post-Cold War national security has been broadened to include those elements that meet a society's personal expectations of the nation's values (Nye 2011, chapter 7; Roskin 1994). The 1997 President's Commission on Critical Infrastructure Protection's definition of national security best captures today's definition of national security: 'confidence that Americans' lives and personal safety, both at home and abroad, are protected and the United States' sovereignty, political freedom, and independence with its values, institutions, and territory intact are maintained' (White House 1997, p. B-1).

4 Geometrically, superposition is one figure in the space occupied by another so that the two figures coincide throughout their whole extent (Webster's College Dictionary 2010) or in quantum physics, in all possible states simultaneously (WhatIs 2015).

5 Computers (or software) are automatically constructing malware independent of human intervention (P.W. Singer in NPR 2013).

6 U.S. Joint Publication 3-13 *Information Operations* considers cyberspace one of 14 different information-related capabilities that contribute to information operations.

7 It is impossible to charge Russia conclusively with orchestrating the attack because of the use of botnets so that the attacks seem to be coming from everywhere, but the available evidence is overwhelming that it was a predesigned Russian atack; the Estonian government originally claimed to trace the source of some of these attacks to Russian governmental addresses. Nevertheless, the nature of the attacks, Moscow's continued sanctions on Estonia, Russian-demanded revision of Estonian laws of its Russian minorities and continued labelling of Estonia as a Fascist or pro-Fascist regime all lend credence to the allegation that the offensive was initiated and orchestrated by Russia (Blank 2016).

8 While it may indeed be a coincidence that RBN operations have on several occasions coincided with official Russian Federation views and/or actions, it is also likely that the Russian leadership is well aware of the capabilities RBN offers and utilises them to assist in achieving international Russian strategic objectives (Fiddner 2011, p. 26).

9 'Based on registration data of the C&C servers and numerous artifacts left in executables of the malware, we strongly believe that the attackers have Russian-speaking origins', (Kaspersky report cited in Holden 2013).

10 Putin has admitted that the war in 2008 with Georgia was planned by Moscow since 2006 (Blank 2016). A concise description of the attacks may be found in Grant 2007, pp. 3–9.

11 Jeff Carr, an investigator for the Project Grey Goose, concluded that 'the level of advance preparation and reconnaissance strongly suggests that Russian hackers were primed for the assault by officials within the Russian government' (Krebs 2008).

12 The cyber-attacks began attacking Georgian websites and discussing upcoming military operations weeks before the actual onset of hostilities, even to the point of conducting what appeared to be a 'dress rehearsal' of the upcoming cyber-attacks (Blank 2016).

13 'All three users who have exploited this vulnerability are very much at risk until they approach us directly to assist with any information. ...The details we have on the hacker are below. Stop at nothing to bring this person to your own definition of justice' (Silk Road 2014).

References

Air Force Research Laboratory. 2003, *Information Assurance Cyber Ecology*. Final Technical Report AFRL-IF-RS-TR-2003-1, January, Rome, NY: Air Force Research Laboratory, Information Directorate, Rome Research Site.

Albert, R., Jeong, H., and Barabási, A.-L. 2000, 'Error and Attack Tolerance of Complex Networks', *Nature*, no. 406, July 27.

Baudoin, C. R. 2013, 'The New Cloud Ecology', Cutter Blog. Cutter Consortium, 18 June, viewed 18 June 2013, http://blog.cutter.com/2013/06/18/the-new-cloud-ecology/.

Blank, S. 2016, 'Information Warfare A la Russe', in Ph. Williams and D. Fiddner (eds), *Cyberspace: Malevolent Actors, Criminal Opportunities, and Strategic Competition*. Carlisle, PA: Strategic Studies Institute, U.S. Army War College, 205–272.

Broad, W. J., Markoff, J., and Sanger, D. E. 2011, 'Israel Tests on Worm Called Crucial in Iran Nuclear Delay', *The New York Times*, 15 January.

Brodie, B. 1959, *Strategy in the Missile Age*. Santa Monica, CA: The RAND Corporation.

Choucri, N. 2012, *Cyberpolitics in International Relations*. Cambridge, MA: MIT Press.

Clark, D. 2010, 'Characterizing Cyberspace: Past, Present and Future', MIT CSAIL Version 1.2, viewed 18 June 2013, https://projects.csail.mit.edu/ecir/wiki/images/7/77/Clark_Characterizing_cyberspace_1-2r.pdf.

Crumrine, C. 2008, 'Anonymous: Cybervigilantes?', Internet Entrepreneurship Blog, 10 March, viewed 15 January 2018, http://internet-entrepreneurship.com/anonymous-cyber-vigilantes.

David, J. E. 2014, 'Head of Online Currency Exchange Found Dead in Singapore', *NBC Business News*, 5 March, viewed 10 June 2014, www.nbcnews.com/business/business-news/head-online-currency-exchange-found-dead-singapore-n45101.

Department of Homeland Security. 2011, *Enabling Distributed Security in Cyberspace: Building a Healthy and Resilient Cyber Ecosystem with Automated Collective Action*. Washington, DC: Department of Homeland Security.

Fausset, R. 2014, Mexican Vigilante Groups Refuse to Lay down Arms in Michoacan', *Los Angeles Times*, 20 January, viewed 10 June 2014, http://beta.latimes.com/world/worldnow/la-fg-wn-mexico-vigilante-groups-michoacan-20140120-story.html.

Fiddner, D. 2011, 'National Cyber Security Strategy Against Malevolent Use of the Global Cyberspace', paper presented to the World International Studies Committee (WISC) 3rd Global International Studies Conference, University of Porto, Porto, Portugal, 17–20 August.

Fiddner, D. 2015, *Defining a Framework for Decision Making in Cyberspace*. Strengthening Cybersecurity Series Report, Washington, DC: IBM Center for The Business of Government.

Frankfort-Nachmias, C. and Nachmias, D. 2008, *Research Methods in the Social Sciences*, 7th edition. New York, NY: Worth Publishers.

Grant, R. 2007, *Victory in Cyberspace*. Washington, DC: US Air Force Association.

Herrera-Flanigan, J. 2001, 'Cybersecurity Ecosystem: The Future? Nextgov: Cybersecurity Report', 24 March, viewed 5 August 2011, www.nextgov.com/cybersecurity/cybersecurity-report/2011/03/cybersecurity-ecosystem-the-future/54390/.

Holden, D. 2013, 'Global Espionage Network Hacks Computers, Smart Phones', 13 January, viewed 4 July 2014, www.sv411.com/index.php/2013/01/global-espionage-network-hacks-computers-smart-phones/.

Ida, K. 2016, 'Transnational Organized Crime and Digilantes in the Cybercommons', In Ph. Williams and D. Fiddner (eds), *Cyberspace: Malevolent Actors, Criminal Opportunities, and Strategic Competition*. Carlisle, PA: Strategic Studies Institute, U.S. Army War College, 513–543.

Jenkins, J. 2007, 'Man Trolled the Web for Girls: Cops', *Sun Media*, 7 December, viewed 17 February 2014, http://cnews.canoe.com/CNEWS/Crime/2007/12/07/4712680-sun.html.

Kan, P. 2013a, 'Cyberwar in the Underworld: Anonymous vs. Los Zetas in Mexico', *Yale Journal of International Affairs*, Winter, viewed 4 July 2014, http://yalejournal.org/2013/02/26/cyberwar-in-the-underworld-anonymous-versus-los-zetas-in-mexico/.

Kan, P. 2013b, 'Anonymous vs. Los Zetas: The Revenge of the Hacktivists', *Small Wars Journal*, 27 June, viewed 4 July 2014, http://smallwarsjournal.com/jrnl/art/anonymous-vs-los-zetas-the-revenge-of-the-hacktivists.

Know Your Meme. 2014, 'Operation Darknet', viewed 16 February 2014, http://knowyourmeme.com/memes/events/operation-darknet.

Krebs, B. 2008, 'Report: Russian Hacker Forums Fueled Georgia Cyber attacks', *The Washington Post*, 18 October, viewed 8 June 2014, http://voices.washingtonpost.com/securityfix/2008/10/report_russian_hacker_forums_f.html.

Libicki, M. 2009, *Cyberdeterrence and Cyberwar*. Santa Monica: RAND.

Libicki, M. 2016, 'Reflections on Cyberdeterrence', In Ph. Williams and D. Fiddner (eds), *Cyberspace: Malevolent Actors, Criminal Opportunities, and Strategic Competition*. Carlisle, PA: Strategic Studies Institute, U.S. Army War College, 391–416.

Mandiant. 2013, 'APT1: Exposing One of China's Cyber Espionage Units', February, viewed 30 April 2017, www.fireeye.com/content/dam/fireeye-www/services/pdfs/mandiant-apt1-report.pdf.

Moktadir, T. 2014, 'Bitcoin: Doom or Blessing?' Light Castle Blog, 2 June, viewed 14 May 2017, www.lightcastlebd.com/blog/2014/06/bitcoin-doom-blessing.

NATO. 2013, 'The History of Cyber Attacks - a Timeline', NATO Review, viewed 4 July 2014, www.nato.int/docu/review/2013/Cyber/EN/index.htm.

NPR. 2013, 'Cybersecurity Forces U.S. To Examine Technological Changes', *National Public Radio*, 20 December.

Nye, J. 2011, *The Future of Power*. New York, NY: PublicAffairs.

Perrow, Ch. 1984, *Normal Accidents: Living with High-Risk Technologies*. New York, NY: Basic Books.

Roskin, M. G. 1994, *National Interest: From Abstraction to Strategy*. Carlisle, PA: Strategic Studies Institute. U.S. Army War College.

Silk Road. 2014, 'Part I: Trouble in Paradise. Comments of DefCon, Administrator of Silk Road Exchange after the 2nd Theft of Silk Road', viewed 13 February 2016, http://silkroad5v7dywlc.onion/index.php?topic=25091.msg491029#msg491029.

The Hacker News. 2013, 'Operation Red October: Cyber Espionage Campaign against Many Governments', *The Hacker News*, 24 January, viewed 4 July 2014, http://the-hackernews.com/2013/01/operation-red-october-cyber-espionage.html.

Thomas, T. 2016, 'China's Reconnaissance and System Sabotage Activities: Supporting Information Deterrence', In Ph. Williams and D. Fiddner (eds), *Cyberspace: Malevolent Actors, Criminal Opportunities, and Strategic Competition*. Carlisle, PA: Strategic Studies Institute, U.S. Army War College, 173–204.

U.S. Chairman of the Joint Chiefs of Staff. 2012, *Information Operations*. Washington, DC: Joint Publication 3-13, 27 November.

Webster's College Dictionary. 2010, *Random House Webster's College Dictionary*. New York, NY: Random House Reference.

WhatIs. 2015, 'Superposition', viewed 16 June 2015, http://whatis.techtarget.com/definition/superposition.

White House. 1997, *Critical Foundations: Protecting America's Infrastructures*. Report of The President's Commission on Critical Infrastructure Protection. Washington, DC: The White House, June.

Wikipedia. 2014, 'Timeline of Events Associated with Anonymous: Chris Forcand Arrest, viewed 17 February 2014, http://en.wikipedia.org/wiki/Timeline_of_events_associated_with_Anonymous.

Williams, Ph. 2011, 'Strategy for Infrastructure Protection and Crisis Management in the Cyber Age: An Elusive Quest?' In D. Caleta and P. Shemella (eds), *Counterterrorism Challenges Regarding the Process of Critical Infrastructure Protection*. Ljubljana and Monterey, CA: Institute for Cooperative Security Studies and Center for Civil-Military Relations, 69–78.

15 Conclusions

Protocols of war – dimensions and layers

Paweł Frankowski and Artur Gruszczak

Introduction

In the introductory chapter to this volume we have suggested the analysis of contemporary security studies as a series of complex interconnections, the so-called protocols of war, that shape and structure forms of international and transnational security. To capture the challenges and developments in the area of international security, widely discussed in the subject literature but presented in a rather descriptive and normative manner, we have intended to offer in this volume an analytical framework which encompasses the increasing impact of new technologies, roles performed by combatants and non-state actors in contemporary warfare, variation among legal and would-be-legalised forms of violence, impact of the market economy and differentiation of cultural explanations of ongoing conflicts. We wished to pay due attention to the normative aspects accompanying technological factors and ethical issues. When the protocols of modern war are opened and run on different levels of contemporary conflicts and crises, they do not operate in a legal vacuum, even if sources of their legitimisation are vague, questionable or diffused. Often the states, international organisations or NGOs are unable, and in some cases unwilling, to move forward in their quest for a better understanding of conflicts they are engaged in. They prefer to neglect the sense and substance of the protocols of war and ignore the whole range of actors involved in hostilities and warfare.

Adopting a market perspective, the protocols of modern war fall into two categories. First, they are supply-driven, i.e. drafted, constituted and implemented both within the state and by the state as well as in the margins of 'official politics', determined by the government and law. They are often written by private hands, thus questioning the state's accountability and responsibility for managing conflict and war (see Hlouchova's chapter). Sometimes they are issued by organised groups challenging the state and undermining international order (chapters by Varin and Thadani). In the most extreme cases, they are brought to the edge of chaos provoked by turbulent transformations triggered by new technologies (the cases of cyber security explored by Burton as well as Fiddner). Narratives present in the supply-driven protocols of war greatly differ in terms of semantics, logic and rationale. Divergent meanings of violence, terrorism, law, morality and justice compete with each other which leads to the cacophony of visions, voices and statements.

Demand-driven protocols are the second category, marked by close, nearly intimate binds connecting war with policy makers and state actors. A strong demand for a novel understanding of contemporary armed conflicts and revisiting the concept of violence results from inability or unwillingness of state actors to redefine violence and warfare against the backdrop of leading protocols of the late 1990s and early 2000s, such as 'post-heroic war' (discussed by Sajduk in this volume), 'post-national war', 'hybrid warfare' or 'humanitarian intervention'. War as a form of political activity needs a rationale justifying the use of violence in the complex, interdependent, networked world of the 21st century. Be this rationale culturally embedded, be it technologically-driven, be it fomented by religious fervour and radical beliefs, it generates strong demand for legitimising adopted views and postures. In this context, the protocols of war seek to meet the demand for an unambiguous, simplified explanation of the complex reality.

We asked the contributors to this volume to (1) identify elements of the protocols of modern war that address specific issues in the areas of the contributors' research and expertise; (2) establish a logical link between the contemporary grammar of warfare and individual actors who frame, implement and develop this grammar; (3) address those aspects of contemporary warfare which exert a particular impact on the public (society/population). Having read the thirteen chapters making up this volume, we have found out that most of the solutions should be looked for in two fields: (1) the contemporary meaning of security and the variety of instruments used for creating and maintaining security; (2) the complex language of war (see Galtung 1987), devoid of clear-cut boundaries and formally rejecting pre-existing terms and notions.

The contributions show that varied groups, states, movements, non-state actors, but also research institutions and academia have made substantial contributions to the contemporary protocols of war. With an advent and expansion of globalisation forces, accompanied by regional idiosyncrasies, there is an urgent need to define boundaries between the scope, meaning, usage and control of the security environment. Since a good number of actors, ranging from the states to private forces and organised mobs, share similar activities undertaken outside the traditionally defined theatres of war, the authors of the present collection have elaborated brilliantly on both singularity and universality of the protocols of war.

The second common thread which has appeared in this volume is the dynamics of the protocols, marked by the increasingly complex and competitive elements of grammar and language of contemporary war and warfare. The criteria measuring success in conflict, prospects for victory, factors contributing to the winning of public support and lessons learned from previous conflicts have to be redefined. The complexity of war and conflict is not a new phenomenon. However, an emphasis must be placed on the changing roles of new actors operating from/within/outside the state. Moreover, it is commonly acknowledged that multi-causality of conflicts and growing interdependence of belligerents result in the fog of war – the Clausewitzian *Nebel des Krieges*. Therefore, capabilities, strategies, plans, rules of engagement as well as generally defined goals and tasks of warring parties cannot be properly understood, described and analysed without an overall assessment of the nature and intensity of risks and hazards. Some authors, instead of emphasising the aforementioned dynamics of protocols, preferred to focus on competition and dichotomy between public and private, modern and traditional ways of conflict and warfare. For others, ethical issues and the logic of power game prevailed. The interplay of traditional threats, instruments of violence and non-traditional aspects of the protocols of war turns us to general findings.

Findings

We have found that the protocols of modern war can be identified in four functional layers: (1) the system (encompassing the art and craft of war, technology, ideology, culture, ethics, law, economy, population), (2) connectivity (interdependent bi- and multilateral linkages binding actors, assets and resources in a networked configuration of the area of conflict and confrontation), (3) communication (channels, codes and tools of distribution and transmission of information, values, norms and rules), and (4) the application (modes and practices of translating the outcomes of war and conflict into political decisions and social behaviour). These layers interact with each other in a way that determines the dynamics of the protocols of war. It is particularly relevant for pointing out the basic features of the protocols and their structural properties of adaptation to the changing face of contemporary conflicts and rivalries.

Protocols of war contain several characteristic elements which put them at the heart of contemporary security studies and need further consideration from social, economic, legal and institutional points of view. The first feature is universality: protocols of war are established and recognised in different areas, regions and regimes. Moreover, they are determined by common technologies and conventional forms of warfare. Loosely organised non-state actors can use modern technologies on an equal footing to state authorities. In terms of confrontation, both can utilise some technologies as a weapon in cyber warfare and do serious collateral harm which may produce considerable devastation of their own assets and resources (see Gross 2010, p. 187). Therefore, the organisation and conduct of contemporary warfare, regardless of the level of societal development, regional conditions and the sources of conflict and grievance, basically follow the similar patterns established on the layers of application and connectivity.

The second characteristic aspect is identification. The protocols of modern war create globally identified elements of the grammar of war (see Butt et al. 2004), whenever new wars, postmodern warfare, hybrid wars or asymmetric conflicts can be observed through the same lens. Paradoxically, war as an instrument of international politics forbidden by international law has experienced a dynamic evolution and adjustment to the changing rules of competition for global, regional and domestic power. As a result, the protocols of war reflected the ways, means and instrument of applying violence and coercion to contemporary politics. The identification and decoding of a given protocol of war often helped to distinguish the state of war of the state of non-war defined as 'hybrid peace' (Richmond 2015).

The third trait is exposure. The protocols of modern war are largely accessible although not easily conceivable. They are not a part of secret operations or intelligence activities; to the contrary, in most cases they are officially declared and pronounced, seeking to project a specific vision onto the layers of communication and connectivity. They blend open access codes with sets of universal messages, commonly shared values and elements of global consciousness. Quite often they seek to garner praise for a comprehensive solution of dilemmas or problems they specifically address.

The fourth feature is apparent fuzziness (see Dimitrov 2003, 2005; Dimitrov and Hodge 2002). The protocols of modern war are somehow 'cursed' by fuzziness and determinism embedded in feigned random behaviour of contemporary belligerents. Therefore, the systemic functional layer is constantly permeated by uncertainty and bafflement as to ontological security and firmness of the elements of the system.

The contributors to this book also sought to figure out how the protocols of modern war, once established, may acquire a political meaning, and to what extent decisions made under a given protocol can have important public policy consequences beyond the realm of confrontation. New methods of warfare, technologisation of war and virtualisation of the realm of confrontation between state and non-state actors raise questions about public awareness and ethical sensitivity in contemporary wars. Taking into account the traditional legal and institutional patterns of global security based on UN Security Council as 'a clearing house and legitimator for a global collective security regime' (Buzan 1991, p. 442), the protocols of war could be conceived as a formula of escapism from a strategic stalemate and fruitless political debates on ethical and legal issues of contemporary warfare.

The protocols of modern war are largely politicised. They frame the concepts of warfare and conflict and shape decisions taken by policymakers with regard to protection of human rights, role of non-combatants, national security guidelines, economic strategies and regulatory behaviour. Whereas protocols of war have broad political and economic implications, one should not forget who teaches the grammar of war, especially when private actors have been engaged in the formulation and implementation of public policies. The chapters in this volume show that the protocols of modern war involve different concepts and models of armed conflicts and warfare. As far as major powers are involved, protocols of war gain importance in terms of military security and modern strategy (see Gray 1999).

The protocols of modern war vary considerably according to instruments of warfare, technological solutions and ethical aspects of contemporary conflicts. Managing today's conflicts is definitely a far cry from implementing the traditional legal and political instruments. Modern conflict management tends towards the blurring of boundaries between combatants, civilians, security forces, national guards, militias, paramilitary groups and hackers. All are considered 'unpriviledged combatants' or 'belligerents' and as such may become targets, means and instruments of armed confrontation. The distinction between combatants and civilians, strongly anchored in international law in armed conflict, is dubious at present. Hence, the outburst and escalation of modern war is less determined by the laws of armed conflict. Mass surveillance, interrogation, targeted killing, information war or cyber-attacks, although causing considerable moral and ethical concerns, are considered indispensable for ensuring the defeat of an enemy all the more so because they are not strictly forbidden by international law. The armed forces which have to comply with the law of international armed conflict, including during crisis management, humanitarian intervention or post-conflict stabilisation, must learn a new grammar of war, and overcome emerging problems not by capacity building but by regulatory solutions. Therefore, humanitarian tasks, based on compliance with a code of conduct, are as important as functional purposes, i.e. the maintenance of law and order when the warring parties or adversaries hold different values and principles. This also constitutes a considerable challenge for the building of domestic and international public support for peaceful solutions to protracted conflicts and long-standing hostilities.

The protocols of war challenge contemporary warfare as a predominantly state activity. Since states have long been the principal actors on the scene of international security, language and grammar of war have been constructed primarily by them. Universality of rules and principles of war and peace, measured by the number of states acceding to international treaties and conventions and adhering to their provisions, placed emphasis on the states acting individually or in international organisations. What has changed, it is the shifting boundaries between the areas of state activities: political, military, technological, ethical, economic, etc., which modify the essence of interactions between state actors. This raises a practical question of the origins and nature of boundaries separating those spheres of state activities. What is more, boundaries translate into organisational patterns which help to understand the protocols of war. One can distinguish spatial, cognitive and temporal dimensions of the boundaries between state domains.

First, spatial boundaries concern changes in territory, geography and space as security variables. They are factors which influence the way contemporary decision makers perceive, experience, understand and recognise the space and territory. This is particularly important for the making of boundaries between cyberspace and physical territory but also no less relevant for hybridisation of warfare especially when territorial jurisdiction of the state is directly and bluntly questioned by belligerents.

The cognitive dimension helps to understand 'war messages' and distinguish 'war instruments' from 'speech acts' underpinning such protocols as 'war on terror', 'war on drugs' or 'holy jihad against infidels'. It concerns not only discursive practices, but also traditional identities and security patterns embedded in societies. Traditional media outlets, often aligned with the state in 'objective' coverage of the events, are confronted with rapidly expanding social media reaching the masses and mobilising them for different types of actions. Cognitive skills and abilities are indispensable for an accurate navigation through state domains and outer areas which are separated by blurring boundaries.

Finally, the temporal dimension reflects intense debate on the perception of time in the context of political decisions, military orders and social reactions to violence and war. New technologies often trigger vehement reactions to acceleration of social, cultural and economic processes. The need for rapid reaction, a key element of the new ways of warfare, illustrates the relevance of the time factor in planning, building and maintaining contemporary defence capabilities. Quick transfers of assets between state domains give the state considerable leverage in implementing effectively its policies and optimising the use of public resources. It also makes the protocols of modern war seek a just-in-time application, 'goodness of fit' on time scales.

The above described boundaries make political actors tend to walk a tightrope, behave with deliberate fuzziness, disregard the lack of support and constantly deny any personal involvement in warmongering. The empirical chapters in this volume on Ukraine (Garrett) and private security (Hlouchova) offer persuasive arguments. In the past, the states were the main originators and enforcers of international law in armed conflict. Nowadays, the governments and public actors abstain from taking full responsibility for softening of clear-cut and coherent legal norms and ethical standards. Traditional instruments of the suppression of violence enlisted in international law, such as sanctions, regional security initiatives or armed intervention, seem to be obsolete when obstinate.

Future research guidelines

The variety and variability of the protocols of modern war should encourage scholars to reframe research on contemporary security in a cross-disciplinary manner. By focusing on narrow, sometimes technical issues, such as military procurement, arms transfers or IT technologies, some questions will continue to await a complex explanation. Post-truth narratives and the power of social media are other incentives to shift the focus of research on the language and grammar of war, so important in analysing hybrid conflicts, post-modern warfare or cyber war.

The authors of this volume have provided ample evidences of complexity and contradiction in contemporary wars and conflicts. Some followed the traditional way of explaining contemporary challenges (Talentino et al.), while others highlighted technology and artificial intelligence as key factors determining the nature of war (Pawinski, Sajduk). A good number of the contributors have emphasised the role of ethics, norms and justice (Varin, Matheswaran, Soare, Thadani) as fundamental terms in the dictionary of protocols of war.

As of 2017, three challenges seem to be emphasised in this volume. First and foremost, terrorist threats have been accompanied by widening inequalities in the developing countries (Varin, Thadani). These threats can be perceived in opposition to global public goods, arising from insufficient commitments of the developed countries to overcome structural asymmetries. Those asymmetries, deeply rooted in the 19th-century legacy and post-colonial malaise, cause global tremors. Although such challenges as terrorism or prolonged conflicts are most severe in the developing countries, recipes and solutions are coming mostly from the developed countries (Varin, Talentino et al.).

The second set of challenges result from the constant technological progress which is accompanied by the decline of domestic and international actors, such as states, regional organisations or the United Nations, to deal effectively with new forms of conflict and warfare. In contemporary conflicts traditional instruments of war (physical force, non-lethal coercion and lethal weapons) are complemented with information and communication technologies generating cyber space utilised as the fifth domain of warfare (Matheswaran). Not surprisingly, the very notion 'protocols' has been borrowed from IT vocabulary. The growing role of autonomous systems, both in cyber space and in the battlefield (Sajduk, Burton) means that the states, perhaps unintentionally, steadily limit their capacity to cooperate in cyber space on grounds of agreed rules and procedures (Fiddner). With the emerging awareness of the need for global cyber security rules and procedures, supported both by private entities and governmental actors, the lack of political support in the UN Security Council illustrates controversies over the making of cyberspace safer and more predictable.

Finally, the advent of postmodern and hybrid conflicts, with post-truth, deception and relativism of such terms as attack, support, rally, belligerent, soldier and force, has resulted in new forms of fault lines and multiple allegiances. Therefore, semantics of modern conflicts (Gruszczak, Garret) implies the need to cooperate on establishing new meanings and notions applied in contemporary wars. Multiple narratives on the wars in Ukraine and Syria are the most telling examples of changes in the dictionary of war. When states, media and politicians are unable to find proper words to determine what kind of situation they are in, the protocols of war may clarify the general picture of war and conflict and give a fuller understanding of causes, results and possible scenarios.

References

Butt, D. G., Lukin, A. and Matthiessen, Ch. M. I. M. 2004, 'Grammar–The First Covert Operation of War', *Discourse & Society*, 15(2–3): 267–290.

Buzan, B. 1991, 'New Patterns of Global Security in the Twenty-First Century', *International Affairs*, 67(3): 431–451.

Dimitrov, V. 2003, 'Fuzziology: A Study of Fuzziness of Human Knowing and Being', *Kybernetes*, 32(4): 491–510.

Dimitrov, V. 2005, *Introduction to Fuzziology: Study of Fuzziness of Human Knowing*. Morrisville, NC: Lulu Press.

Dimitrov, V. and Hodge, B. 2002, *Social Fuzziology: Study of Fuzziness of Social Complexity*. New York, NY: Springer Verlag.

Galtung, J. 1987, 'Language and War: Is There a Connection?', *Current Research on Peace and Violence*, 10(1): 2–6.

Gray, C. S. 1999, *Modern Strategy*. Oxford: Oxford University Press.

Gross, M. L. 2010, *Moral Dilemmas of Modern War. Torture, Assassination, and Blackmail in an Age of Asymmetric Conflict*. Cambridge: Cambridge University Press.

Richmond, O. P. 2015, 'The Dilemmas of a Hybrid Peace: Negative or Positive?', *Cooperation and Conflict*, 50(1): 50–68.

Index